xGENIUS

JAMES TIPPETT

xGENIUS

EXPECTED GOALS AND THE SCIENCE OF WINNING FOOTBALL MATCHES

BLOOMSBURY SPORT
LONDON · OXFORD · NEW YORK · NEW DELHI · SYDNEY

BLOOMSBURY SPORT
Bloomsbury Publishing Plc
50 Bedford Square, London, WC1B 3DP, UK
29 Earlsfort Terrace, Dublin 2, Ireland

BLOOMSBURY, BLOOMSBURY SPORT and the Diana logo are trademarks of
Bloomsbury Publishing Plc

First published in Great Britain 2024

A catalogue record for this book is available from the British Library

Library of Congress Cataloguing-in-Publication data has been applied for

ISBN: TPB: 978-1-3994-1155-4; eBook: 978-1-3994-1156-1

2 4 6 8 10 9 7 5 3 1

Typeset in Adobe Garamond Pro by Deanta Global Publishing Services, Chennai, India
Printed and bound in Great Britain by CPI Group (UK) Ltd, Croydon CR0 4YY

To find out more about our authors and books visit www.bloomsbury.com
and sign up for our newsletters

In loving memory of conventional football statistics.
xG is here now.

CONTENTS

FOREWORD

It is my pleasure to introduce you to *xGenius*, an exploration of how xG and other advanced metrics are changing the sport of football. Given my role as a football pundit, I'm a big advocate of xG in that it gives me a greater understanding of who has dominated a game rather than just looking at the scoreline or who had the most shots.

Introducing xG to a wider television audience hasn't always been easy. We definitely had to push for a while to get it into the statistics at the end of a game, especially on Super Sunday. There was plenty of pushback from other pundits, although I won't name names! We've been using xG and other similar statistics on Monday Night Football for a while now, and it's really added a layer of depth to our analysis.

Analytics is always progressing, and I love that. I started playing around the same time as Prozone launched, meaning we had access to statistics like how far we had run. The level of insight is in a whole new world these days. I'd have loved to have seen the data we use today attached to my game, and where I would sit in terms of the modern-day players.

In my job, it's vital you keep up to date with modern technology and data. That doesn't mean you have to embrace everything, but I think you have to try to understand it and delve into why people are using it. What are they looking to get out of it?

In this book, James dives deep into the world of football statistics and explores insights, trends and stories that will change the way you watch football. Whether you're a coach looking to improve strategy, a fantasy football enthusiast looking for an edge or simply a fan wanting to be more knowledgeable about the game, this book has something for everyone. It is an essential read for anyone who wants to understand more about the beautiful game. Enjoy!

Jamie Carragher
Football pundit and author of *The Greatest Games*

INTRODUCTION

On Christmas Day 1914, British and German troops laid down their arms, climbed out of their trenches and played a game of football in no-man's land. They set aside their animosity on that cold winter morning, greeting each other with goodwill while exchanging gifts. Amidst the snow, they set up goals using empty cans and helmets. The soldiers played, oblivious to the danger that surrounded them, united by a shared passion for the beautiful game. This story depicts a temporary respite from the horrors of the First World War, a brief moment of humanity and solidarity that transcended the boundaries of nationality and ideology.

But nothing is mentioned of who won the game. There is no match report we can refer to, no website which provided live updates of the score. Similarly, we don't have any information on the performance of either side. There weren't any Opta analysts present at the game to log the number of passes, tackles or shots. Perhaps the British soldiers laid siege to the Germans, creating an endless string of chances but failing to put the ball in the makeshift goal. Perhaps the Germans scored with their only chance of the match – a flank guard operating as an overlapping wing-back putting in a cross which was skilfully headed home by a brigadier from 12 yards out. Perhaps, as they trudged back to their trenches after the game, the British soldiers bemoaned their poor finishing ability and rued their bad luck. Perhaps they assigned each of the shots which took place in the match a percentage chance of resulting in a goal. Theoretically, that match had an xG scoreline.

On an autumn day in 1946, thousands of football enthusiasts gathered to watch the opening ceremony of the Hackney Marshes football pitches in East London. It was a momentous occasion in the history of grassroots football, the unveiling of a patch of land which would become the spiritual home of Sunday League football. The occasion was marked by the unfurling of a giant Union Jack flag and the singing of the national anthem. As the ceremony drew to a close, players from local teams stepped out onto the fields for the first time. The opening games were played with the same passion and vigour as the hundreds of thousands of Hackney Marsh matches they preceded. These matches each had an xG scoreline.

The Rwandan Genocide of 1994 left the country devastated and divided. In the aftermath of the conflict, football provided a rare opportunity for people from different backgrounds to come together, share their love of the game and build bridges across ethnic and political divides. Through football, Rwandans were able to put aside their differences and work towards a shared vision of a brighter future. In July 2003, the Rwanda men's national team beat heavy favourites Ghana to secure their qualification to the African Cup of Nations finals for the first time in their history. The final score was 1-0, but as both Tutsi and Hutus danced together in the streets of Kigali, the result was not as important as the message it sent: that Rwanda was ready to move forward and build a better future for all its citizens. That historic match had an xG scoreline.

The 2005 Champions League final was a game for the ages. Held at the Atatürk Olympic Stadium in Istanbul, the match was contested between AC Milan and Liverpool. Milan, boasting a star-studded line-up including the likes of Andriy Shevchenko and Paolo Maldini, were heavy favourites going into the game. The first half was a disaster for Liverpool as their Italian opponents raced into a 3-0 lead, seemingly putting the game to bed. The Milanese were playing with a level of skill and intensity that seemed almost superhuman. But then something remarkable happened. Liverpool, led by captain Steven Gerrard, scored three goals in six second-half minutes to level the score. The match went to penalties, where Jerzy Dudek saved Shevchenko's spot-kick to complete one of the greatest comebacks in footballing history. That match had an xG scoreline.

Every match in the history of football has had an Expected Goals scoreline. Well, sort of. They all had shots which could be assigned a percentage chance of being scored. We might adapt the age-old question of whether a fallen tree makes a sound if no one is around to hear it to the following: 'If a football match happens without an analyst present, does it have an xG score?' We know the Expected Goals scorelines for top-level games played after xG data started being widely collected in the mid-2010s. Analysts have re-watched and collected Expected Goals for some games further back, but we'll never know the xG data for most of the games in the history of football. The stats for the Christmas Day truce of 1914 will forever be out of reach.

The reader has a similarly unknown cumulative xG total from all the football matches they've played throughout their life. Think of every shot you've ever taken, whether in the playground as a child, at five-a-side or in Sunday League football. These shots all carried a probability of being scored based on their location, the foot you used to strike the ball, the number of defenders in the way and so on and so forth. Add all these probabilities together and you have a life-long cumulative xG score. Just because you don't know what it is, doesn't mean it doesn't exist. Expected Goals is, in a way, inevitable and inescapable.

For years, fans had a sort of subconscious Expected Goals model operating inside their heads. When a game ends, you naturally reflect on who should have won. This opinion is based on the scoring opportunities presented to either side. *'If we had scored that penalty we would have won.' 'They were so lucky to have scored that deflected shot from long-range.' 'We created so many good chances, how did we lose?'* The subconscious opinion of whether or not our team deserved to win isn't an exact figure. It's a feeling. The Expected Goals method simply seeks to put a numerical value on the intuitive instinct fans have harboured for decades. It's a currency which imbues shots with value. A penalty is worth 0.77(xG), a long-shot might be worth 0.03(xG). Earning as much of this currency as possible will give teams the greatest chance of success.

The intuitive xG model living inside each of our brains taps into our earliest primordial instincts. Human beings are controlled by algorithms made up of emotions, sensations, and thoughts. The exact same kind of algorithms control baboons, chickens, otters, and pigs. Consider the following survival problem: a baboon spots some bananas hanging from a tree, but also notices a lion lurking nearby. Should the baboon risk its life for the bananas?

This boils down to a mathematical problem of calculating probabilities: the probability the baboon will die of hunger if it does not eat the bananas versus the probability the baboon will be caught by the lion. The baboon needs to process a lot of data in order to reach a decision. How far away are the bananas? Are there impeding obstacles in the way? Where is the lion positioned? Does the lion seem alert? Does the lion seem hungry?

Earning as much xG currency as possible will give teams the greatest chance of success.

This act of processing a series of data points to determine the probability of an outcome occurring is a lot like the act of working out the xG of a shot. Instead of lion, read goalkeeper; and instead of obstacles, read defenders. Of course this is all very abstract. How exactly does the baboon calculate the probabilities? It certainly doesn't draw a pen from behind its ear, a notepad from its back pocket, and plug the numerous variables into a sophisticated 'Expected Bananas' model. Rather, the baboon's entire body is the calculator. What we call sensations and emotions are in fact algorithms. Baboons and humans alike are attuned to measuring the *feeling of danger* involved in a certain situation. This extends to chance creation in football. When a big scoring opportunity arises at a football match, you often hear a collective cry go up inside the stadium. That is a natural reaction to a dangerous situation. It's animalistic fear. Chance creation in football accesses that prehistoric area of the brain that detects panic and makes us feel threatened. The striker bearing down on our goal might as well be the lion taking a step towards us as we dash for the bananas. Our brain responds: Danger! Our hearts start pumping faster and our hands start sweating. That's what Expected Goals measures: danger. It's a mathematical representation of the threat posed to our (or, hopefully, our opposition's) goal. This is even conveyed in the language we use to describe football. Skilful attacking players are a 'threat', while a team who create a lot of chances are 'dangerous'. The dialect we attach to chance creation is steeped in the innately human emotion of fear.

For those who need a refresher, Expected Goals, otherwise known as xG, is the stat which lends its name to the title of this book (well, sort of). In a low scoring sport such as football, the chances that a team create and concede are a better indication of their ability than the number of goals they score. Goals are incredibly random; chance creation is more systematic. Every shot which takes place in a match carries an xG value – the percentage chance of that effort resulting in a goal. Several different factors impact the xG value of a shot, including the distance and angle to goal, whether the shooter used their strong foot, weak foot or head, the number of defenders in the way, the type of attack the chance has stemmed from, and a number of other variables depending on what model you use. Adding up the xG of each shot from either team will leave you with an overall xG scoreline for the match, such as **AC Milan (2.68) 3-3 (2.14) Liverpool.** In this game, the aforementioned 2005 Champions League final, Milan can feel slightly aggrieved not to have lifted the trophy at full time given the quality of chances they created and conceded.

A frequent misconception with xG is that it attempts to predict how many goals will be scored in the future. The name 'Expected Goals' leads people to think the metric tells us how many goals we can expect a team or player to score in an upcoming game. This is not the case. Expected Goals looks back in time, rather than forward in time, and tells us how many goals a team would have expected to score based on the opportunities they created in the past. Of course, this information gives us a strong insight into the ability of teams and, as a result, can help us better forecast how well they will perform in the future, but the statistic is primarily designed to help us judge past performances. If we could rename Expected Goals, it might be better off labelled 'Chance Quality', 'Goal Probability', or something similar.

Note that the xG value doesn't take into account the quality of the shooter. A common question is asked: 'Surely, all other things being equal, a shot from Harry Kane is worth more than a shot from Harry Maguire?' This may be the case, but Expected Goals values every shot as if it was taken by the average player. For example, a penalty is worth 0.77(xG) regardless of who takes it. There are several reasons for this. First, taking the probability of an 'average' player scoring allows us to measure over- and under-performance. If Harry Kane consistently outperforms Expected Goals, scoring more than the xG model expects him to, then we can say he's either a good finisher or has been lucky. Second, finishing ability is incredibly inconsistent. Even the greatest players go through long periods of xG underperformance. How are we supposed to factor in the quality of the finisher when scoring output is so erratic? If Harry Kane scores 0 goals from 4.00(xG) over several games, how should this impact how we measure his 'finishing ability'? And what timeframe should we measure this supposed metric over? In fact, the degree of variance in the xG overperformance of individual players has led many to question the extent to which 'clinicalness' even exists. But we're getting ahead of ourselves. The crucial thing to note at this point is that xG measures every shot as if the average player was taking it.

The Expected Goals Philosophy, the predecessor to this book, aimed to bring xG to the attention of the wider football community. It sought to explain how xG worked and where it originated from, as well as providing some high-level examples of how it can be applied. *xGenius* will take a deeper, more forensic look into how the lens of football analytics is becoming sharper, using real-life examples to explore the benefits of the xG way of thinking. 'Analytics is not done with statistics,' Bill James, the father of baseball's sabermetrics movement, once said. 'It's done by applying scientific methods to real-life problems.' This book will put xG and other revolutionary metrics into action and demonstrate how several clubs, companies and individuals are using analytics to unlock the science of football. If *The Expected Goals Philosophy* described how to build a Formula 1 car, *xGenius* unleashes it on the track.

Chapter 1 will introduce the clubs at the forefront of the football analytics revolution. These teams are the pioneers who have harnessed the power of xG and advanced analytics to propel themselves up the football league ladder. They've managed to turn over hundreds of millions of pounds of profit in the transfer market and have turned conventional football wisdom on its head. These clubs will serve as regular case studies throughout this book.

Chapter 2 explores how we can even begin to make sense of a sport as messy and noisy as football. The amount of luck that resonates throughout each match makes it difficult to understand what's really going on. How can we differentiate between what is signal and what is noise, what is truth and what is fiction?

Chapter 3 pivots into the television studio, putting pundits under the microscope. These 'experts' are celebrities in their own right, with millions of people tuning in each week to listen to their insight and analysis. But how well do pundits actually understand what it takes to win matches?

Chapter 4 transports us to the trading floors of some of the most secretive and covert betting operations in the world. These companies fully understand the science of what makes teams win football matches and have been able to make vast fortunes by predicting the outcomes of games.

Chapter 5 is when we first peek behind the veil at professional football clubs. We start at the top, with the owners and the boardrooms. How have innovative leaders structured their organisations to give their teams the best chance of on-field success?

Chapter 6 knocks on the door of the manager's office. How important is the man in charge, and how should we assess managers in the modern day? The story of football's analytical revolution is, in part, the story of a struggle between analysts and management staff.

Chapter 7 journeys into the recruitment department of football teams. This is perhaps the most crucial cog in the machine of a club. Every season we see teams of players battle it out on the field, when secretly the real battle is taking place between analysts, scouts, and data scientists to identify the best talent and bring them to their clubs.

Chapter 8 explains how xG has altered the way teams play the sport. The realisation that not all shots are created equal has led managers to evolve tactics in the pursuit of 'big chance creation'. What systems, formations and routines have smart clubs developed in order to create more xG, score more goals and collect more points?

Chapter 9 explores the mythical 'position of maximum opportunity' – or POMO, for short. Where do goals actually come from? Since the advent of xG, the average shot distances in Europe's major leagues have plummeted season-upon-season. The pursuit of the optimum, highest-value way of playing football means counterintuitive measures must be taken.

Chapter 10 investigates the world of set pieces. Dead-ball situations are the low-hanging fruit which analytical clubs have grabbed with both hands. Mastering the art of set pieces can improve your goal difference by 20 goals each season. It is often the difference between winning and losing.

Chapter 11 takes a closer look at the dynamics of 'finishing ability'. Do players actually possess a God-given talent at finding the back of the net, or do some simply benefit from the favour of Lady Luck?

Chapter 12 examines a deeper analytical toolkit which we can use to illuminate the beautiful game. Expected Goals has become the poster boy for football analysis, but there are several other metrics worth exploring if we're going to develop a full understanding of the science of winning matches.

We have data on hundreds of thousands of shots, each one a unique experiment into how football works.

Chapter 13 ventures into the dressing room. In a game so increasingly focused on individual talent, the output of teams is often ignored. Can we quantify the impact of team chemistry? And can we measure how well all 11 players on a field are synchronised in their movements, their ability to create space for one another? This is perhaps where the science of football is at its most advanced.

Chapter 14, the final section, reflects on how the tidal wave of xG has swept over the shores of football in recent years, as well as looking towards the future of sports analytics. The same philosophy which has brought football teams success is now being applied to other games. Golf, basketball, cricket, hockey, tennis and many other pastimes have adopted an xG way of thinking which is helping coaches and players win more often.

A decade ago, this book couldn't have been written. Although many of the questions posed – those of finishing ability, playing style and transfer strategy – have been around for decades, we've only recently unearthed the data required to provide the answers. Chance creation has always been at the centre of football. Journalists wrote match reports centred around the key goalscoring opportunities at either end of the pitch long before the advent of three-minute-long YouTube highlight packages. But the invisible hand of xG has only recently revealed itself in the flesh. Now we are able to count and quantify the ability of teams to create dangerous situations. We have data on hundreds of thousands of shots, each one a unique experiment into how football works. This has completely transformed the landscape in terms of how we evaluate and think about footballing performance. Other books on this subject have told the story of human beings who use football statistics. This book aims to make the data itself the central character. Hopefully *xGenius* offers the reader a new perspective, like wearing x-ray glasses to reveal the hidden structures and forces behind the chaotic and messy reality of the sport.

As well as answering some of the puzzles we've been trying to solve for generations – 'how good actually is my team?', 'where do goals come from?', 'what shots are most valuable?' – the dawn of the xG era has also introduced new ones – 'does finishing ability exist?', 'how much impact does game state have?', 'what balance should clubs strike between data-driven methods and traditional ways of thinking?'

Football is 'the infinite game.' No one will ever be able to fully understand all its nuances, and no one will ever be able to 'master it' or 'complete it.' All we can do is get as close to the truth as possible. Think of each football match as an image. Looking at it simply through the lens of the final scoreline means the image is incredibly blurry. When you begin to layer in conventional stats – possession, shots, shots on target – you begin to improve the focus. Introducing the sort of advanced metrics we'll study in this book – Expected Goals, Field Tilt, Expected Possession Value – will provide even more clarity. The aim of football analytics is to make the image as high-resolution as possible. This book explores the interplay between analysis, tactics, and decision making. It seeks to put the sport under the microscope with the aim of getting closer to the ultimate truth of what makes players, managers, and teams successful. What, ultimately, wins football matches.

1

THE MODEL FOOTBALL CLUB
THE TEAMS THAT CHANGED THE GAME

'Innovation, not increased funding, can be the only route to success for clubs such as ours'

Matthew Benham, Brentford FC owner

Graham Potter puffed out his cheeks and tilted his head to the sky as Christian Benteke wheeled away in celebration. Brighton & Hove Albion had just conceded a 95th-minute goal to go 2-1 down against their bitterest rivals. Potter was having one of those nights that every football fan has endured at some point. His team had dominated the game, taking 25 shots to Palace's three. They had completed 13 passes within 20 yards of their opponent's goal, conceding zero of such passes. Brighton had created far better scoring opportunities, accumulating 2.06(xG) to Palace's 0.24(xG). If the exact same match was played out one hundred times, Brighton would have won on 93 occasions and lost just once.[1]

Brighton travelled to West Bromwich Albion five days later. Kyle Bartley headed the hosts in front after 10 minutes with an effort that could generously be described as a half chance. Brighton laid siege to their hosts over the course of the next 80 minutes, missing two penalties and a host of other big chances. The match finished in a 1-0 defeat, despite an Expected Goals scoreline that read 3.14(xG) to 1.13(xG) in favour of Graham Potter's men. For the second consecutive match, the chance of Brighton losing based on the game's scoring opportunities rounded to 1 per cent. For the second consecutive match, that 1 per cent likelihood took place.

A week later, Brighton lost their third match in a row. A tightly contested game once again saw the gods rule against them as they fell to a 2-1 defeat to Leicester City. The loss left them 17th in the Premier League table, clear of the relegation zone only by virtue of a superior goal difference to Fulham, albeit having played one game fewer. At this point in the season, early March 2021, Brighton had played 27 matches. They'd scored 16 goals fewer than expected and conceded five goals more than expected, according to xG. They'd accumulated 26 points, but leading Expected Goals models indicated they should have amassed roughly 46 points based on the quality of their performances throughout the campaign. Brighton would have occupied fourth place if the table were ranked according to Expected Points, behind only Manchester City, Chelsea, and Liverpool.

Brighton's extraordinary run of form throughout the 2020/21 season can be described using a host of adjectives. Unlucky. Implausible. Brutal. But overall, it was *funny*. It was funny in the ridiculousness of it, in the absurdity of it. We often make exaggerated claims about sport having the ability to write scripts that no human could conjure up, but surely even the most imaginatively cruel author couldn't invoke the tale of Brighton's woes that campaign. They were playing

---

[1] That isn't just rhetoric. If the match was simulated thousands of times according to the xG figures at the end of the game, Brighton would have won in 93 per cent of instances and Crystal Palace just 1 per cent.

teams off the park and creating an enormous number of chances but ending up on the wrong side of the result time and time again.

This tragicomedy playing out on the Premier League stage caught the imagination of social media. One Twitter account (as it was then known) was perfectly placed to tell the narrative of Brighton's tale of anguish. 'The xG Philosophy' (handle of @ xGPhilosophy) had started posting post-match Expected Goals scorelines when football returned from the Covid break. The first significant Brighton result of note came in late September 2020, when they lost 3-2 to Manchester United despite creating 3.03(xG) to United's 1.91(xG). (This game also famously saw United awarded a penalty via VAR after the full-time whistle had blown, which Bruno Fernandes duly converted to win the match in the 100th minute). From there, the cult perception of Brighton as 'the xG team' gained increasing momentum with each undeserved defeat. Memes were shared which jeered Brighton's demise and gained huge traction across 'football Twitter'.

The perception of Brighton as 'the xG team' gained increasing momentum with each undeserved defeat.

This saga had wider consequences than simply the mockery of a south-coast football club. There was still an inherent distrust of xG from the wider football community at this point in the evolution of football analytics. The comments under every @xGPhilosophy post at the beginning of the 2020/21 season skewed more towards rejection than they did acceptance, but Brighton's performances managed to turn that tide. The non-believers were watching Brighton's matches, seeing their incredible dominance, then having visibility of the xG stats after the game. The data was validating the eye test. Brighton's results weren't aligning with their performances and the xG deniers began to understand what the metric was all about.

Engagement began to snowball. Starting from an initial base of 8,000 followers in June 2020, @xGPhilosophy grew to over 200,000 in the space of a year. The new followership included notable figures within the world of punditry such as Jamie Carragher, David Jones, Micah Richards, Michael Owen, and Xabi Alonso. The former two of these began integrating xG into Monday Night Football, Sky Sports' leading football analysis show. The craze even sparked the attention of Brighton's official Twitter account, which took notice of the cult following their team had built in this space and began self-deprecatingly replying to @xGPhilosophy's posts on a regular basis.

Brighton did end up avoiding the cold grasp of relegation that season, finishing in 16th position on 41 points. Based on their xG performances, leading models estimated they should have finished on 62 points, the fifth-best Expected Points total of any team in the league. Their underperformance versus expectation was the most severe of any team since xG data has become available, and the most severe we will likely see for quite some time. The cruel irony of this story is that Brighton are pioneers in the global football analytics revolution. They were, in fact, one of the first teams to harness the power of xG.

CLIMBING THE PYRAMID

Rewind back to before the turn of the millennium: Brighton were a club in dire straits. Having lost the Second Division play-off final in 1991, the club entered free fall. Two relegations later, The Seagulls found themselves needing a win against Hereford United on the final day of the 1996–1997 season to avoid relegation to non-league. A late equaliser from Robbie Reinelt ensured Brighton retained their league status by the tightest of margins. The turn of the millennium saw Brighton forced to move to Withdean Stadium, a converted local athletics track which was owned by the council. The capacity of the ground was a measly 6,000 initially, increasing to 8,500 after a couple of years. It was voted the fourth-worst football stadium in the UK by *The Observer* in 2004. The temporary nature of the ground was obvious. It was primarily used for athletics, with a single permanent stand along the north side and the remainder of the stands assembled from scaffolding. Changing and hospitality facilities were provided by portable cabins placed haphazardly around the site. The ground summed up Brighton's status as a football club in the early 21st century. This unhappy stage of their history was spent bouncing around the lower echelons of the Football League.

But one day, everything changed.

Tony Bloom allegedly placed his first bet at the age of eight, when he would use his pocket money to play the fruit machines at the nearby arcade. Bloom's passion for gambling continued well into his teenage years. At the age of 15, he regularly snuck to the nearest city to make use of a fake ID for betting purposes. He went on to study Statistics and Mathematics at The University of Manchester, before securing a job as an accountant upon graduation. His career path led him from the trading room floors of the city to those of the bookmaker Victor Chandler (which would later become BetVictor). Here, Bloom learned his craft, acquiring a profound understanding of the betting markets while also tinkering with his own model for predicting the outcome of football matches. He eventually went out on his own, setting up several online gambling sites during the early 2000s and occasionally taking time out to play poker against the world's elite. Bloom's first major win came in 2004 when, at the age of 33, he won the Australasian Poker Championship in Melbourne, collecting the first prize of £180,000. By 2008, his live tournament winnings exceeded £1,200,000. His cold-blooded nature, expressionless stare and cool decision-making bestowed upon him a nickname: 'The Lizard'.

Bloom drew inspiration from this nickname when founding a betting consultancy, Starlizard, in 2006. The company's success was based on Bloom's top-secret model for predicting match outcomes. In simple terms, they were first to the xG party and managed to drink all the free booze before the rest of the guests turned up. Starlizard used, and continues to use, Expected Goals data to form accurate gauges of a team's ability. Their xG data allows them to identify undervalued selections in the betting markets and has enabled them to make hundreds of millions of pounds in profit. By 2009, Bloom had acquired enough capital to swoop in and rescue his childhood football club, the struggling Brighton & Hove Albion. Bloom applied the same statistical analysis that brought him success in the world of betting to his running of Brighton. Expected Goals data informed Brighton's decision-making across several verticals; from recruitment, to style of play, to opposition scouting.

When Bloom took over in May 2009, Brighton were about to embark on their fourth consecutive season in League One, the third tier of the English pyramid. A couple of years later, in their final season at Withdean Stadium, the club secured promotion to the Championship. They moved to a brand-new ground called Falmer Stadium (later renamed the Amex for sponsorship purposes) and duly changed their crest to a design similar to that used from the 1970s to the

1990s, reflecting the club's return home having not had a stadium of their own since 1997. The rise did not stop there. Shrewd recruitment, a restructuring of the management staff and the channelling of statistical analysis allowed the club to defy the odds and achieve promotion to the Premier League for the first time in 2017, just two decades after they were on the brink of elimination from the Football League (although they played in the top flight from 1979 to 1983, when it was known as the First Division). Since then, Brighton have been a mainstay in the richest division in the world, despite possessing a budget which is dwarfed by many of their competitors. Their infamous 2020/21 xG car crash of a campaign was followed up with their highest ever league finish in 2022, ninth in the Premier League. In 2022/23 they managed to achieve a European spot by finishing sixth, despite possessing the third-lowest wage budget in the league.

Brighton aren't the only club operating in the analytical half-space. Although all Premier League teams now possess some form of analysis department, three English clubs stand apart from the rest in their wholesale adoption of analytics. Of the triumvirate, Brentford FC are the most similar to Brighton in their structure and approach. In fact, the two teams are football's equivalent of identical twins. Both clubs are run by a fan who made their millions by applying a rigorous statistical approach to the betting markets. They've both adopted an xG-driven approach to running their clubs and have structured their organisations to tailor for a 'data first' philosophy. Both clubs have enjoyed an incredible rise up the Football League pyramid as a result of outthinking, rather than outspending, the opposition (*see* Figure 1.1).

For Tony Bloom, read Matthew Benham. For Starlizard, read Smartodds. Bloom and Benham were initially colleagues at an online bookmaker that Bloom founded called Premier Bet. Benham worked under Bloom, but the pair had an acrimonious falling out. After leaving Premier Bet, Benham founded Smartodds in 2004. Two years later, Starlizard came to be. The offices of the two companies are just down the road from one another, in Kentish Town and Camden respectively. North London has a strong claim of being the Wall Street of professional football betting.

Matthew Benham's football club has followed a remarkably similar rise as Tony Bloom's. Brentford were a mid-table League One team when Benham

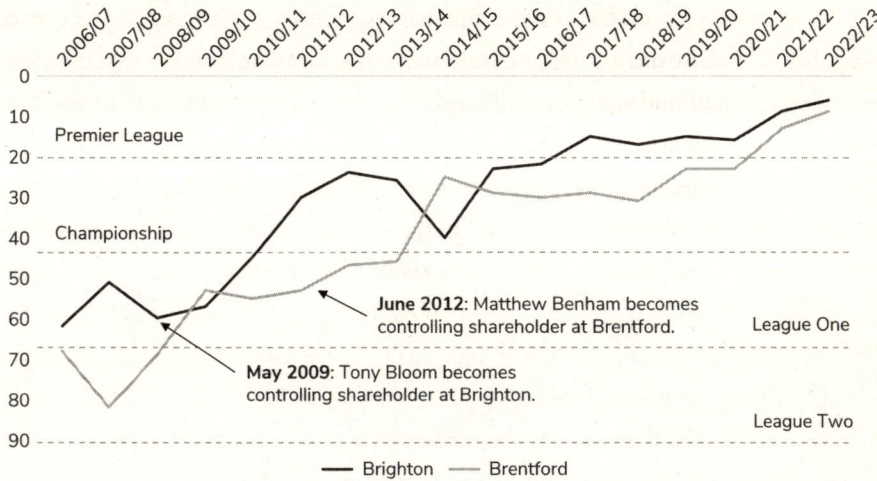

Figure 1.1: Brentford and Brighton League Finishing Positions, English Football League 2006–2023

took the reins in 2012. Two years later, they were promoted to the Championship for the first time in 20 years. Despite possessing the fourth-smallest playing budget in the 24-team division and being heavily tipped for immediate relegation, the club finished fifth in their first season in the second tier. After a few years of consolidation at that level, Brentford reached the top division for the first time since 1947. Their first season in the top flight saw them finish 13th in the table, despite being the poorest club in the league. The team then defied all logic by achieving a top half finish in 2022/23. Brentford have been transformed from a mid-table third division team into a comfortable Premier League club in less than a decade. An embracing of analytics appears to carry an anti-gravitational, upwards force that shoots your team to the top of the league pyramid.

Wage expenditure has long since been recognised as the strongest indicator of a team's performance. The deeper the owner's pockets, the better players their team can attract, meaning the higher up the league table they finish. In 2022/23, Brighton had the second-lowest annual wage bill of the 20 Premier League clubs. Despite this, their performances on the pitch led to a sixth-place finish and a Europa League spot. Their xG performances were even better. According to Expected Points, Brighton should have finished in the top four positions and earned a Champions League spot. Although possessing the wage budget of a relegation-threatened team, they were actually the fourth-best team in the league that season according to Expected Points.[2]

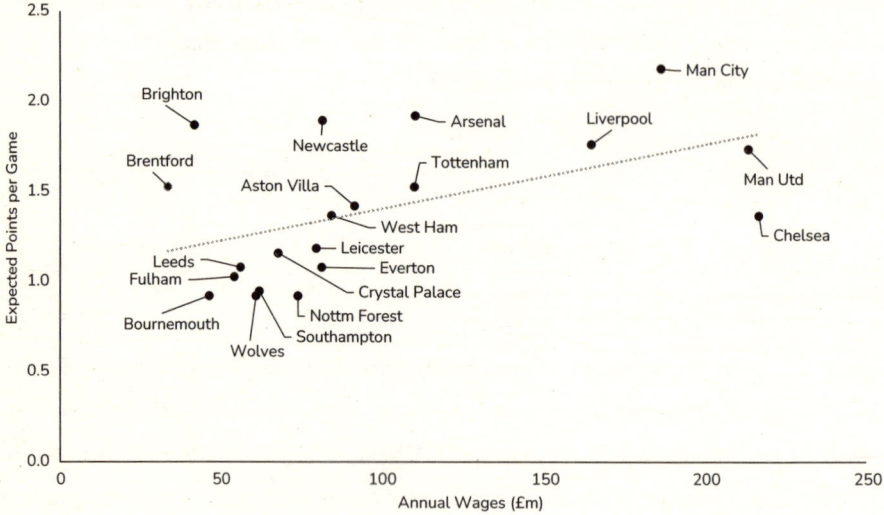

Figure 1.2: Expected Points per Game vs Annual Wages[3], Premier League 2022/23

Brentford were equally as impressive, with an annual wage bill of just £34 million. According to every rule of football logic, the Bees should have finished in last place. Instead, they narrowly missed out on a Europa Conference League

---

[2] We will study Expected Points later on. In simple terms, it tells you how many points a team deserved to take from a game based on their xG performances. Note that Expected Points doesn't try to tell us how many points a team were expected to achieve before the game took place, but rather tells us how many they deserved to win based on how well they played (in the same manner as how xG tells us how many they should have scored based on the chances they created). A team who dominate a game in terms of chance creation but end up losing will amass close to 3 Expected Points, while a team who get an incredibly lucky win will get close to 0 Expected Points. Adding up a team's figures over the course of a season will give an alternative, more truthful representation of their standing.

[3] Although different sources give slightly different figures of Premier League wage expenditure, the same trend prevails: Brighton and Brentford were the two poorest teams in the division.

spot on the final day of the season. They beat treble-winning Manchester City twice, the only team to do the double over Pep Guardiola's side that season. They also defeated Manchester United, Liverpool, Tottenham Hotspur, and Chelsea, meaning they had the best record against the 'big 6' teams of any side that season. Expected Points revealed them as the seventh-best team that campaign and worthy of securing European football.

Team	Annual Wages (£m)	Expected Points	xPts per £m
Arsenal	110	73	0.66
Aston Villa	92	54	0.59
Bournemouth	47	35	0.75
Brentford	**34**	**58**	1.73
Brighton	**42**	**71**	1.70
Chelsea	217	52	0.24
Crystal Palace	68	44	0.65
Everton	82	41	0.50
Fulham	54	39	0.72
Leeds	56	41	0.73
Leicester	80	45	0.56
Liverpool	165	67	0.41
Man City	186	83	0.45
Man Utd	213	66	0.31
Newcastle	82	72	0.88
Nottm Forest	74	35	0.47
Southampton	62	36	0.58
Tottenham	110	58	0.53
West Ham	85	52	0.62
Wolves	61	35	0.57

Figure 1.3: Expected Points per Wage Expenditure, Premier League 2022/23

Unsurprisingly, Brentford and Brighton come out as the best teams in terms of Expected Points per £1 million spent on wages during 2022/23. While it's true that this view does penalise the big clubs for spending several times more than the smaller clubs, the gulf in financial efficiency between Brentford and Brighton and the rest of the division is undeniable. Based on wage spend, the strongest predictor of success in football, both teams should have been relegated to the Championship. In reality, their xG-driven philosophies allowed them to thrive in the best league in the world. This book will explain how Brighton and Brentford have used data to turn conventional football logic on its head.

These teams are the poster boys of football's analytical revolution, and other clubs are sitting up and taking notice of their methods. In May 2023, Tottenham chairman Daniel Levy began sounding out betting experts in an attempt to replicate the Brighton-Brentford model. Levy realised that Spurs were struggling to keep pace with other big Premier League clubs, while the two data-driven teams were able to achieve sixth- and ninth-place finishes respectively despite having considerably shallower pockets. Brighton and Brentford's success was born from the betting markets; their owners' skill at analysing football and predicting future outcomes. Both clubs live and breathe the science outlined in this book and, as such, will feature regularly as case studies. Their journeys weren't always smooth. From terrible runs of form to dramatic clashes between the analytics department and management staff, all against the background of immense budgetary pressure, these two Davids have managed to take on and defeat entire leagues worth of Goliaths.

The third club in the triad of analytical frontrunners could certainly be described as one of these giants. Liverpool are European heavyweights who have used data analysis to achieve both domestic and cross-continental glory in recent years. They have faced a different set of challenges to Brighton and Brentford. The latter clubs were in dire straits before being rescued by lifelong fans in Bloom and Benham. Bold and innovative strategies are far easier to deploy at the bottom of the ladder. Beggars can't be choosers, but football royalty certainly can be. A club the size of Liverpool, despite operating on a smaller budget than certain other teams in England and across Europe, will naturally find it more difficult to achieve buy-in for new or risky ways of thinking. The Reds offer an interesting case study when examining how to implement the science of winning matches at an elite club, and one we'll visit periodically throughout the book.

XG DISCREPANCIES

Critics of xG might argue that Brighton's near-relegation in 2020/21 contradicts the argument that data analysis should be a cornerstone of each football club's operations. The Seagulls represented an xG believer's perfect blueprint for how a team should be run. They possessed a forward-thinking manager who played an attacking brand of football, an innovatively-structured front office, armed with the best data and analytical toolset anywhere in the world, a highly-skilled management staff packed with a diverse range of experiences, perspectives and ways of thinking, and a recruitment team with a proven track record of mining the transfer market and identifying high-quality players at bargain prices. It all came together on the pitch, as the team created great chances and prevented their opponents from doing so. Yet they still almost got relegated simply because a handful of opposition goalkeepers played out of their skin, some penalty decisions went against them, and their strikers miskicked the ball more often than usual. This team became the laughingstock of the footballing world. Why bother to innovate and forward-think when freakish bad luck means even the best-run clubs can struggle in any given season?

Another data sceptic might use Brighton's infamous season as ammunition to launch a different attack on Expected Goals. '*When the xG is so drastically different from the actual number of goals a team scores, it shows how useless and inaccurate Expected Goals is as a metric,*' such a critic might shout. Condemnation of this sort came to the fore during the 2022 World Cup. Only 15 of the first 32 matches (47 per cent) at the tournament were won by the team who created the higher xG figure. In other words, the team who created the most xG failed to win more often than not. A handful of noticeable xG scorelines captured the attention of xG deniers, notably:

Argentina (2.27) 1-2 (0.14) Saudi Arabia

Germany (3.53) 1-2 (1.33) Japan

Belgium (0.86) 1-0 (2.83) Canada

'*Surely,*' the cynics argued, '*this disproves xG as a valid statistic.*' Other sceptics claimed that these scorelines proved the model was flawed. Ironically, the results experienced in the above matches and throughout Brighton's 2020/21 season are the situations when we need our xG x-ray glasses the most. The critics are missing the point. The whole purpose of xG is to highlight when performances aren't matching expectation. In the above matches, the xG isn't differing from the goals. *The goals are differing from the xG.* Expected Goals shows the actual ability of a team; the final score is the product once you add a whole bunch of randomness and noise into the mix.

The purpose of xG is to highlight when performances aren't matching expectation.

Suppose you decide to throw a normal six-sided dice 100 times and count up the total number of 'points' that are accumulated. Before the experiment, you might expect an overall points total of 350, given that the 'average' score of a dice throw is 3.5[4]. Say your 100 rolls actually produces a score of 500. Does that mean your dice is broken? Not necessarily. It just means your 100 rolls threw up a series of unusually high numbers. This is useful information. In a football match, both teams are attempting to increase the number of rolls (shots) that they're allowed, while also increasing the numbers on each side of the dice (the quality of their shots). But those teams who manage to succeed in both tasks aren't necessarily going to end the game with the highest points total (number of goals).

There have been other, more valid criticisms of xG, and @xGPhilosophy's Twitter/X presence in particular. The account is set up to post xG scorelines in the following format: **Brighton (2.06) 1-2 (0.24) Crystal Palace**, with the bracketed numbers representing the xG. The benefit of this format, compared to different accounts who post xG shot maps, timelines, and other graphics from matches, is simplicity. The account's success has stemmed from the adage, 'less is more'. The xG stats are presented in an instantly digestible manner and can be easily consumed without having to open an image or read through numerous lines of text. This has facilitated widespread engagement, which in turn has

---

[4] 3.5 is the average of the following list of numbers: 1, 2, 3 ,4, 5, 6.

helped xG to reach a wider audience and permeate mainstream football consciousness.

The downside of @xGPhilosophy's format is the lack of context, which the audience is expected to provide for themselves. It doesn't show whether there was a penalty in the game, nor whether a team played some of the match with 10 men. Game state can also play a factor in the final xG totals – a team who scores early might sit deep and defend in an attempt to protect their lead. These are important bits of information that cannot be captured in a simple xG scoreline, just as they cannot be captured in the usual post-match presentation of a scoreline (e.g. Brighton 1-2 Crystal Palace). Data alone is inevitably limited in its capacity to describe the world. The application of context is crucial.

Each Expected Goals figure isn't reality. It's an attempt at representing reality to the best of our ability. The US federal reserve defines a mathematical model as 'a representation of some aspect of the world which is based on simplifying assumptions'. Essentially, a phenomenon will be represented mathematically in order to produce a simplified 'pretend' version of reality. Each shot's xG number signifies an attempt to define the chance of that shot resulting in a goal given the outcome of hundreds of thousands of previous shots with similar characteristics. Different models churn out different xG figures for the same shot, but the differences are usually minimal and over the long run all models tend to more or less align. Think of xG models like judges at a boxing bout. One may score a shot one way and another a different way due to their own perspectives. They all have strong points and blind spots. Back in football, one model may occasionally underrate the value of big scoring opportunities, another might not take into account the position of the goalkeeper. Usually, the xG models agree on the winner of a game, but sometimes it comes down to the equivalent of a majority or a split decision.

The reason xG models don't always agree with one another in the short-term, and the reason xG can never completely mirror reality, is because they offer non-verifiable judgements. Consider a prediction being made about the peak temperature in Barcelona tomorrow, or the result of a presidential election. If you disagree with a friend about these questions, you will, at some point, find out who is right. Now consider a company deciding which of two candidates to hire as CEO. Whichever choice they make, they'll never truly be certain they made the right one. They can never be certain that the other option would or wouldn't have performed better than the candidate they chose. Similarly, if an event which a forecaster assigned a 90 per cent chance of occurring fails to happen, the judgement of probability wasn't necessarily a bad one. After all, outcomes that have a 10 per cent likelihood of occurring do occur 10 per cent of the time. Any probability judgement (i.e., anything between 0 per cent and 100 per cent) can never be truly confirmed or denied. Expected Goals data is an example of such non-verifiable judgements. We can say a shot has a 30 per cent of hitting the back of the net, but we will never be able to verify this in the same manner as we can verify the peak temperature in Barcelona tomorrow.

CHAPTER SUMMARY

- Brighton's 2020/21 season was one of the unluckiest in recent football history. The team consistently played well but struggled to turn chances into goals and good performances into points on the board.
- The Expected Points per Game vs Weekly Wages graph illustrates how well Brighton and Brentford have been performing on the pitch relative to their wealth, validating their use of xG in everything from recruitment to opposition analysis.
- A wider context is required, above and beyond xG, to understand how a game has developed. Game state often plays a critical part in adding context to the final xG totals – and that is information that cannot typically be used to capture an xG scoreline.
- xG has moved into the mainstream thanks to a growing knowledge of Brighton and Brentford's use of it for recruitment, aligned with increased online interest in the idea thanks to vehicles such as the @xGPhilosophy Twitter/X page.

2

THE FIELD OF PLAY

HOW RANDOMNESS INTERACTS WITH FOOTBALL

'The only sure thing about luck is that it will change'

Bret Harte, 19th century American poet and author

Sergio Agüero. Troy Deeney. Marcello Trotta. These three players have few, if any, noticeable similarities. They each hail from a different corner of the planet; Argentina, England, and Italy respectively. They've played football at many different levels and in different leagues, from the Argentine Primera Division, to Serie C, to the Midland Football Combination Division Two. You've almost certainly heard of one of these players, probably heard of two of them, but almost certainly not heard of all three. The only substantial connection to be made between the triad is their participation in a last-minute, season-defining moment.

On the last day of the 2011/12 Premier League season, with the score tied at 2-2 and Manchester City needing a win against Queen's Park Rangers to secure the Premier League title, Agüero collected the ball just outside the box and drove towards goal. He cut past a defender before unleashing a powerful right-footed shot that flew into the back of the net, sparking wild celebrations at the Etihad Stadium and securing City's first top-flight title in 44 years. The goal was a pivotal moment in the club's history and the defining moment of Agüero's career.

The next season, Vicarage Road staged one of the most extraordinary play-off moments in history. With Watford and Leicester City tied 2-2 on aggregate and the game entering the final seconds, Anthony Knockaert won a soft penalty for the visitors. A goal would have sent Leicester to the final, but Manuel Almunia intervened by saving both Knockaert's penalty and the resulting rebound. Vicarage Road erupted as the ball found its way to the feet of Fernando Forestieri on the right-wing. The Watford player surged upfield and crossed into the box, whereupon a knock-down was smashed home by Troy Deeney. Pandemonium ensued. The footballing world had never seen anything like it before. Although that wasn't quite true. Two weeks prior, a remarkably similar and similarly remarkable incident had taken place just across the capital.

In the final game of the League One season, Brentford needed to defeat Doncaster Rovers to achieve promotion to the second division of English football for the first time since 1993 and the second time in 50 years. Anything less than a Brentford win would see Doncaster promoted instead. With the score tied at 0-0 heading into the 95th minute, the Bees won a penalty. Up stepped Marcello Trotta, a 20-year-old Italian on-loan from Fulham. Twelve thousand fans held their breath inside the packed Griffin Park. Many turned away, not daring to watch. The referee blew his whistle. Trotta ran forward and sent the ball high to Neil Sullivan's right. The goalkeeper watched as it smashed hard into the underside of the crossbar and bounced down back into the penalty area. There was a momentary melee in the Rovers box, before the ball was eventually cleared. Brentford had committed players forward in an attempt to scramble home the rebound, which allowed Doncaster to run straight down the other end and score.

It was one of the most dramatic endings to a league season in football history. Doncaster had been one kick away from heading into the play-offs in third position, but 20 seconds later had secured the title. As for Brentford, they ended up losing the League One play-off final to Yeovil Town a few weeks later. For the Bees fans, it felt like a pivotal moment in the club's history. The team had blown their best chance of promotion to the Championship for two decades. Little did they know, a far more pivotal moment had happened a couple of weeks prior at the club's training ground, Jersey Road. This moment was a first meeting between Matthew Benham and Rasmus Ankersen, whose union would not only end up changing the trajectory of Brentford FC, but fundamentally alter the way the sport of football is played.

A GAME OF CHANCE

Rasmus Ankersen grew up as one of Denmark's most promising young footballers before a knee injury on his first ever senior appearance effectively ended his playing career. Despite, or perhaps because, his own gift had been snatched from his hands, Ankersen developed a deep interest in talent, particularly in where it comes from and how to identify it. He gained his UEFA A License and became the assistant coach at FC Midtjylland U17s, the same team he played for at youth level. The club lacked the financial might or locational appeal to draw in high-quality players and so relied heavily on producing young talent. Ankersen gained exposure to one of the best football academies in Europe and a club with a strong culture for innovation. He drew upon this experience to write a book entitled *DNA of a Winner*. The book sold well enough for him to quit his job at Midtjylland, move to Copenhagen and start working as a performance coach. There, he wrote two more books, *Leadership DNA* and *The Goldmine Effect*. The latter of these caught the attention of Matthew Benham, the owner of Brentford. Benham's club lacked the stadium infrastructure or fanbase to make money from ticket sales. The club had little commercial appeal, meaning lucrative sponsorship deals weren't an option. And the television income for League One teams was next to nothing. The only way Brentford could generate enough revenue to cover their costs was through the smart trading of players. Benham realised Ankersen's expertise in the identification and nurturing of talent could prove a useful asset to the West London club.

At the time of their meeting, a few weeks prior to Trotta's penalty miss, Brentford were fourth in League One with five matches remaining, though with games in hand on the teams above. Ankersen asked Benham, 'What do you think? Are you guys going to be promoted?' Benham gave a response that Ankersen didn't quite expect. It wasn't an answer full of excitement, nervousness, or any other glimmer of emotion. He simply looked at him and said, 'At the moment, there's a 42 per cent chance we'll be promoted.' In that moment, Ankersen realised he'd met someone who thought completely differently about the game of football than anyone he'd ever met. Benham and Ankersen met regularly over the next couple of months to exchange ideas about how a football club could be run differently. They were particularly interested in how it would be possible to break the strong correlation that existed between spending and results. A team's wage bill is the strongest indicator of their success and Brentford's pockets were shallower than those of their opponents. The two men embarked on a mission to outthink, rather than outspend, the competition. Their weapons of choice in this battle were data, analytics, and a profound understanding of the laws of probability.

Benham and Ankersen appreciated how difficult it is to analyse the sport of football. The main problem is that goals are an incredibly scarce resource. On average, there is a score every nine minutes in American football, every 13 minutes in rugby and every 22 minutes in hockey. Football, on the other hand, sees a goal scored every 35 minutes. Around 8 per cent of matches finish goalless, while 40 per cent of games are decided by a one-goal margin. A goal is celebrated in football with a degree of ecstasy that no other sport can parallel. Spectators are made to wait for the ultimate reward, the reward that wins matches and rockets players to stardom. However, the infrequency with which goals are scored is problematic for those trying to make sense of the game.

Commentators often remark, 'At the end of the day there's only one stat that counts. How many goals you score'. Thousands of actions take place each match: passes, tackles, shots, duels, sprints and so on. However, at the end of the game these events are rendered inconsequential. Law 10, as written by the International Football Association Board, is the law that governs all others; 'The team scoring the greater number of goals is the winner. If both teams score no goals or an equal number of goals the match is drawn.' The only thing that matters in football is how many times a team manages to hit the back of the net, and how well they prevent their opposition from doing so.

The dynamic between the incredible significance and complete unpredictability of goals is what makes football so entertaining. The Sergio Agüero and Troy Deeney last-gasp winners and Marcello Trotta's penalty miss serve to outline the fine lines which exist between success and failure. The beautiful game is gilded with randomness. One kick of the ball can decide an entire season. The journeying up and down the country for nine months, the 40-odd matches, the money the fans spend on travel, tickets, and subsistence – the outcome of it all can come down to the width of a crossbar in the 95th minute of the final game. When there is so little to choose from between victory and defeat, promotion and relegation, glory and failure, the role of luck becomes pivotal.

Goals incorporate a huge amount of chance in terms of how they are scored. If the foot of Agüero, Deeney or Trotta had struck the ball just a couple of centimetres off-kilter, the fate of their clubs' entire seasons would have been drastically different. Every goal in history has had a series of events align perfectly for the scoring team, whether that be defensive error, attacking brilliance or a combination of the two. Fans will often look back at the goals their team concede and regret the slight mispositioning of their right-back, the flappy hands of their goalkeeper or the clinical finishing of the opposition striker. Each of these occurrences will be deemed as 'unfortunate' or 'bad luck.'

The timing of goals is also distributed randomly. Consider two teams in the same league who each play 38 matches in a season. Suppose Team A scores 76 goals and concedes 38 (meaning a +38 goal difference), while Team B scores 57 and concedes 38 (meaning a +19 goal difference). The table shows that, over the 3,420 minutes each team played, Team A performed better than Team B by a goal differential of +19. However, suppose Team A's goals were distributed in the manner that saw them dominate a few games (winning 4-0, 5-1, and so on) but fall to a lot of close defeats (0-1, 1-2, etc.). Team B could feasibly end up with more points because their goals were distributed in a more favourable fashion.

The same principle is true in sports like tennis or darts, where matches are broken down into sets, games, and points. Imagine a tennis player who wins several games 40-0 but then loses slightly more after deuce. This player could feasibly win a significantly greater number of points than their opponent, but end up losing the match because their counterpart's points fall in a convenient order. Similarly, while a 0-3 and 1-2 defeat in football both return the same number of points (zero), a team losing by tighter margins is probably better. For this reason, goal difference generally speaks more to the quality of a football team than their points total does. Football analysts will often use the GD column as a barometer of team skill before they look at the points column. The latter is subject to the inconsistency and randomness of how goals distribute themselves, while the former is a net representation of how that team has performed throughout the season as if they'd played one big 3,420-minute match.[5]

How and when goals are scored can be described as incredibly *noisy*.

How and when goals are scored can be described as incredibly *noisy*. The term 'noise' has come to represent unexplained variability within a data sample. In this context, the term originated from signal processing where it was used to refer to unwanted electrical or electromagnetic energy that damaged the quality of signals and data. The presence of noise is an irritant for those trying to predict repeatable outcomes. Noisy data is data that's lost meaning by the existence of too much variation and randomness. In football the signal (the true quality of teams and players) is obscured by noise (the randomness with which goals are distributed and the luck that generates them).

Noise is prevalent everywhere we look. When presented with the same patient, doctors will often make different assessments of their ailment. When presented with the same exam paper, teachers will often give different marks. When presented with the same candidate, interviewers will often give wildly different reviews of their quality. In football, two exact same teams can play each other 10 times in a row and produce several different scorelines and results.

---

[5] Of course, some teams may prove exceptions to this rule. A team might be particularly good at holding on to one-goal leads, meaning their goal difference isn't as flattering as their ability suggests.

Noise creates problems for those trying to define the truth and reach a correct or fair outcome. Suppose someone is convicted of a crime – drink-driving, assault, burglary or so on. What is the sentence likely to be? The answer shouldn't depend on the particular judge to whom the case has been assigned. It would be unfair for three people who have committed similar offences to receive drastically different penalties: probation for one, two years in prison for another, and five years in prison for another. And yet this injustice can be found in many countries around the world – not only historically but also in the modern era. The beautiful game puts football teams on trial and Lady Luck is the judge who hands out the sentences. Football analysts grapple with this, trying to work out what swayed her decision and what *should* have happened given the case (quality of performance) made by either team.

Of course, it's not just the randomness of goals that causes noise in football. Referee decisions are inherently inconsistent and generate a large amount of unpredictability. Two identical fouls might induce significantly different reactions from a referee. The governing bodies have attempted to cancel out some of this noise by introducing a Video Assistant Referee system (VAR). Allowing video replays offers the officials the chance to make a second judgement and (in theory) should lead to fewer mistakes and more consistent decisions. In practice there is still a degree of variability to the rulings being made.

Who is the main opponent that a footballer faces? Is it the opposing 6-foot, 2-inch centre-back tracking his every move? The opposition manager conspiring strategies to nullify him and his teammates? Or the bobble of a ball that makes his touch elude him, the gust of wind that steers his shot off target? The invisible hand of chance that takes matters out of his control? Sport's protagonists voluntarily place themselves into an environment where they are surrounded by luck. Their aim is to overcome their exposure to chance by asserting their superior skill and willpower. Only by exposing themselves to this chance, competing with it and against it, can they ultimately overcome it. Bill James, the baseball revolutionary whose ideas inspired Billy Beane of *Moneyball* fame, wrote, 'the problem is that baseball statistics are not the accomplishments of men against other men, which is what we are in the habit of seeing them as. They are accomplishments of men in combination with their circumstance.' In other words, circumstance and chance are the primary opponents against which sportspeople are battling. The context they find themselves in and the pushing and pulling of luck within that context is crucial to their success.

Each match contains a series of events that are subject to the whims of chance. A team scores an unlikely goal, a referee gives a dubious penalty, a player's rash challenge sees them sent off. All of these things could easily not have happened. Matches are shaped by random occurrences. The paradox of sport is that it condenses all of these random events into a single objective result. It converts indeterminacy into a single and brutal conclusion – the final scoreline. One answer emerges from a cloud of possibilities. When we see Barcelona win a football match, we attach a certain meaning to that result. But what if they hadn't scored that early goal? What if a particular throw-in decision had gone the other way? What if a centre-back had chosen to pass to the right-back instead of the left-back in the fourth minute? We live in a world which could have unfolded in many different ways.

Every sport or game features a tension between the prevalence of luck and skill. Any sport that sits too far to one side of this sliding scale would not be enjoyable. A game where every result was simply decided by a random generator would be incredibly dull. The complete absence of skill does not satiate our competitive spirit, our desire to demonstrate talent and resolve. Conversely, a sport in which the favourites emerge victorious every time would also be tedious. International football outside of major tournaments falls victim to this because the disparity in quality between teams is too great. No one is particularly interested in watching England beat San Marino 10-0. Perhaps domestic football has managed to conquer the world because it hits the sweet spot between predictability and unpredictability. We have a semi-firm grasp on which teams are better than others, but we can be proved wrong in any given match.

Football analytics is our attempt to strengthen our grasp on this knowledge, our endeavour to rid the game of chance and uncover the true skill of teams and players. This first rule of football analysis is perhaps the most important, and certainly the most foundational: we must first acknowledge, accept and embrace the randomness that permeates football. Once you've accounted for and removed the role of chance, what you're left with is pure skill. And skill is what we're interested in. How good is your football club? What about your manager, is he doing a good job? Should your striker be scoring more goals? For over a century these questions have been answered subjectively. We've had a *feeling* of how teams and players have been performing. Performance analysis now offers us the tools to strip noise from the sport and more meaningfully answer these questions. Luck in football is like the weeds in your garden. You can never quite fully be rid of them, but they can be kept at bay by using sound analytical processes and more potent exterminators such as xG and other advanced metrics.

Phil Giles, Brentford's Director of Football and one of the key figures in the club's rise up the league pyramid, was once asked how much of a role randomness plays in football. 'About 80 per cent,' he answered. 'It's a lot higher than people think, how much luck plays a part in shaping results, shaping careers, shaping decisions. So many things need to go right for you to be successful, but so many things can go wrong for it not to work out'. The clubs who have mastered the science of winning football matches started by acknowledging that Lady Luck is in the driver's seat. The question then becomes: how do you seize control of the steering wheel?

CHAPTER SUMMARY

- The low-scoring nature of football means luck and chance play a much larger role than in other sports.
- Good teams often lose and bad teams often win due to the randomness which exists in football – this is why we need our xG x-ray vision to help us work out who's actually playing well.
- The timing distribution of goals is incredibly random, meaning the goal difference column often gives a better insight into team ability than the points column.
- In order to uncover the science of what wins football matches, we must first accept and embrace the luck which exists within the sport.

3

THE TELEVISION STUDIO

PUNDITRY AND THE SCIENCE OF FOOTBALL

'The fox knows many things, but the
hedgehog knows one big thing'

Isaiah Berlin

Most football fans will be unfamiliar with the works of Isaiah Berlin. No, he's not a tricky left-winger who plays in the Bundesliga. He was a social theorist who, in 1953, authored an essay which divided writers and thinkers into two categories: hedgehogs and foxes.

Hedgehogs view the world through the lens of a single defining idea. Berlin noted examples included Plato and Friedrich Nietzsche. More modern examples include Winston Churchill (who didn't let contradictory ideas interfere with his fixations), George Washington (who knew one big thing: that America's future as a nation lay to the West) and Wendy's (the fast-food chain that doubled down on its core burger offering).

Then there are foxes, who draw on a wide variety of experiences and address a wide range of topics. Berlin referenced Aristotle, William Shakespeare, and Leo Tolstoy. Readers might be more familiar with examples like Hillary Clinton (who is comfortable addressing a broad range of topics) or McDonald's (the fast-food chain that diversified its menu and broadened its offering). The terms 'hedgehog' and 'fox' have come to represent opposing definitions of human character. In simple terms, a person can be hedgehog-like or fox-like in their approach to analysis and prediction-making.

Karl Marx, the father of communism, believed that his extreme form of socialism was the inevitable conclusion to human society. He viewed the world in terms of class; there were the rich and exploitative, and there were the poor and exploited. According to Marx, the masses would be forced to suffer under the dominion of the aristocracy, until eventually the working class would revolt and seize control of the state. Eventually all power would erode away, leaving in its place a utopian society where everyone is equal, an idea best encapsulated by arguably his most famous utterance: 'from each according to their ability, to each according to their need.' Despite contemporary critics of Marxism who predicted the theory would prove misguided, the radical socialist stuck to his belief and lived his life by this single, overbearing philosophy. The fact he was unwilling to change his forecast under any circumstance makes Karl Marx the archetypal hedgehog.

A fox-like person, on the other hand, constantly updates and revises their analysis as they go along. This wily creature reacts to changing circumstances as they develop and is cautious in their prediction-making. They think probabilistically and appreciate the role that chance can play in any undecided outcome. Bookmakers are stonewall foxes. They carefully analyse the ability of teams and players to work out their odds of success. Any new information that arises – an injury to a key player, rumours of unrest in the playing camp – is quickly fed into their models. During a match, bookmakers constantly update their predictions based on various events – goals, cards, the scoring chances each team is creating, and so on. If they are not flexible in their

Hedgehogs	Foxes
Less willing to admit they might be wrong	Aware they might not know the full picture
Approach the views of outsiders with scepticism	Open to new ideas from all sources
Rigid in their ways	Adapt to different methods
Adopt an 'all-in' approach to prediction-making	Naturally cautious
Don't look beyond 'Yes' or 'No'	Think probabilistically
Generally more entertaining and controversial	Less bold or outlandish

Figure 3.1: The Differences Between Hedgehogs and Foxes

approach and adapt to changing conditions, their predictions will become stale and they will lose out to the bettors. Perhaps communism would have survived longer if Marx had altered his predictions in a similar manner, adapted to changing political circumstances, or if his thinking had reached a compromise with other ideologies. Karl Marx wouldn't have been a very good bookmaker.

In 1995, Alan Hansen famously told the *Match of the Day* audience that 'you can't win anything with kids' when discussing Manchester United's aspirations that season. Hansen was speaking after Alex Ferguson's side had lost by two goals to Aston Villa on the opening day of the 1995–1996 campaign. The line has gone down in folklore as one of the worst predictions in football history. Manchester United went on to win both the Premier League and FA Cup that season and launched a dynasty that lasted over a decade. Ferguson's team became one of English football's most successful sides, and Hansen's statement became one of English football's worst forecasts.

Hansen demonstrated a classic example of hedgehoggery. A fox would have assessed the situation, perhaps stating that it is *unlikely* that you could win a trophy with such a young team. Hansen tried to make his theory sound more valid by confidently proclaiming that Manchester United had no chance of success. The assurance with which he dismissed Ferguson's side is striking. 'They've got problems,' he said. 'The trick to winning the championship is to have strength in depth and they just haven't got it.' In reality, Hansen surely knew that Ferguson had at least a small chance of winning a trophy that year. Why, then, did he make a prediction he knew would probably fail? Why did he allow himself to act like a hedgehog?

To answer these questions, we must dig deeper into what the role of the pundit actually is. These 'experts' are tasked with generating engaging content for us, the general public, to consume. An audacious prediction-maker is more exciting than a rational one. Having an Alan Hansen on your show makes for more interesting viewing than having some analyst who has worked out the exact probability that Manchester United would win the title. Even if a hedgehog like Hansen makes a prediction which is probably not going to come to fruition, a viewer can take a certain pleasure in disagreeing with the pundit. Fans often criticise managers, claiming that they could do a better job of running the team themselves. The same applies to pundits. Deep down, every supporter believes themselves to be an expert of the sport. Hansen's 'you can't win anything with kids' forecast may not have been a successful prediction, but it did receive a great deal of publicity and has been viewed hundreds of thousands of times online. The BBC and *Match of the Day* won't care that one of their pundits was disastrously wrong, but they will care about the increase in views across their social channels.

Large accounts which are part of 'Football Twitter/X' perfectly demonstrate the 'all-in' nature of such predictions. They'll post forecasts like '*Marcus Rashford is 100 per cent going to score today, mark my words*' or '*Barcelona are going to beat Real Madrid 3-0, save this tweet*'. If their unlikely prediction comes off, then great. They'll milk it for all it's worth. If it goes disastrously wrong? No matter, they'll still have hundreds of replies from opposition fans mocking them. These Twitter/X accounts are often anonymous, represented by a display picture of a footballer from the team they support. They won't be bothered by the mass derision headed their way when they're wrong. In fact, they'll enjoy it. For them, any engagement is good engagement. Some corners of the punditry world follow a similar philosophy. Being wrong or right is an afterthought; the main objective is to stay relevant. This approach contradicts sound analytical practice which aims to define the truth and has no care for acclaim or attention.

This problem stretches beyond football. Open any newspaper or turn on any TV news show and you find experts who forecast what's coming. A few are cautious. Most are bold and confident. A handful claim to have supernatural abilities to foresee years into the future. With few exceptions, they aren't placed in front of the cameras because they possess any proven skill at forecasting. Accuracy is seldom even mentioned. Old predictions are like old news; easily forgotten. Pundits are rarely asked to reconcile what they forecasted with what actually happened. The prevailing, undeniable talent that these commentators have is their ability to craft a compelling story and relay it with conviction.

Old predictions are like old news; easily forgotten. Pundits are rarely asked to reconcile what they forecasted with what actually happened.

Pundits and the media don't have any checks on them. There is no inherent *need* for them to be right. It doesn't matter if a pundit makes a handful of failed predictions – they will still be offered more punditry jobs. Hansen didn't get sacked because of his disastrous forecast. If anything, his increased publicity would have secured him even more job offers.

Football punditry is a fairly trivial exercise compared to, say, that of the political realm. Politics is full of so-called 'experts' who offer strong opinions on the world around them – which candidates will be elected, whether one country will invade another, what policies are likely to be introduced in the near future. These predictions have real-world consequences, but studies tracking the accuracy of these forecasters have shown many bear more resemblance to chimpanzees throwing darts at a dart board than knowledgeable specialists who have a strong authority of their subject matter. Again, these pundits will skew more towards ambitious and outlandish forecasts than measured and considered ones. The analyst who correctly calls a recession that no one else saw coming will gain fame, whereas the analyst who never strays far from consensus will remain obscure. That's not to say that all political pundits are useless. The task of peering into the future and perceiving future outcomes is a difficult one. There are obstacles to foresight that may not be surmountable, and our desire to reach into the future will almost always exceed our grasp. But football, in particular, has fostered a culture where open-mindedness and inquisitiveness is being suffocated under the weight of uninformed jargon. Without these qualities, we'll always struggle to cultivate the requisite forecasting skills.

Roy Keane is a classic hedgehog. The Irishman is not one for sitting on the fence and his famed scowl is enough to cast terror into the hearts of his fellow panellists. Don't expect him to meet you halfway or to assess both sides of an argument. He's more inclined to make bold arguments, annihilating any opponent who dares challenge his point of view. Keane seldom takes into account the role of probability; uncertainty is a weakness which makes it look like you don't know what you're talking about. Cross him at your peril.

The interesting thing about Roy Keane is that he seems to live a dual life. 'Roy Keane the Pundit' has all the qualities outlined above, but 'Roy Keane the Person' seems to carry a completely different presence. Watching him off-screen, in podcasts or features where he's not acting as a pundit, he displays behaviour unrecognisable to the man described above. He's calm and considerate with his choice of words and even exhibits, dare I say it, fox-like tendencies. The curious case of Roy Keane hints at an important truth: the nature of punditry turns foxes into hedgehogs. The viewing audience love 'Roy Keane the Pundit,' not because he's a particularly astute analyst, but because he offers incredible entertainment value. The most viewed videos on the Sky Sports YouTube channel feature vociferous debates between pundits, and Roy Keane is usually involved.

Conflict gets traction. But in pursuing this 'entertainment factor', pundits risk losing the analytical, shrewd elements of their character. Punditry is a game of 'survival of the abrasive.' Just as the theory of evolution dictates that only the fittest survive, punditry tends to weed out moderate opinions and make successful those who are outlandish and strong-willed. This is not necessarily the fault of pundits or the media. We, the consumers, must share the blame. We're the ones relentlessly consuming this form of content – conflict over consideration, argument over analytics.

A BRAVE NEW WORLD

Historically, the hedgehog-like nature of punditry has had implications for the broader football community. The way fans talk about and analyse football stems from the personalities they watch on TV. Pundits dictate the language of football and, in the past, this language has disregarded a smarter dialect driven by data and analytics. However, punditry has come on leaps and bounds in the last few years. Sky Sports' *Monday Night Football* has put xG and other advanced metrics at the centre of the conversation. The BBC were the first movers in this space in the UK when *Match of the Day* started showing xG scorelines after the highlights of games back in 2017. The Athletic are perhaps the leaders when it comes to data journalism in football, and have even hired a team of specialised analytics writers.

The media are certainly getting better at incorporating data into our native vocabulary by blending advanced stats with engaging narratives. In yesteryear, stats like shots, shots on target and possession were the most advanced you'd read about in a national paper or have beamed into your living room by a television. Now, the likes of xG, Field Tilt and Expected Threat – all of which we'll study in more detail in latter sections of this book – are being used in mainstream reportage. High-quality, statistically oriented journalism requires investment: the data costs money, you have to build tools to harbour and present the data, you need specialist staff members and you need to invest time into building a culture and environment where analytics can come to the fore. Outlets like The Athletic have taken the game to the next level, telling engaging stories that are founded on cutting-edge research and data analysis. They've proved that fans are interested in data journalism, and that investment in this remit can reap rewards.

There is greater scrutiny on pundits nowadays, which in turn has forged better analysis. Their role used to be to provide colour to games, and while that is still the case to some degree, there is now a greater emphasis on providing insight. Pundits nowadays get deeper into the tactical side of the game; they delve into the actual performances of the players. Their aim should be to replicate the sort of insight that a manager or an analytics department will be going into inside clubs. That's the bar. Studying the little things, taking the scientific examination of the game to the next level, putting players under the microscope, carrying out a forensic examination of each match. They should be using xG to see how well clubs are playing, then cross-reference that with tactical insights to see why teams are performing the way they are.

Not only have the media become better at integrating stats into their broadcasts, they've started using visually aesthetic graphics to engage their audience. The presentation of xG has been key to its wider adoption. Analysts have taken inspiration from an unlikely source in their visual representations of xG data. In drama, three unities represent the Aristotelian theory of dramatic tragedy. There is the unity of place, which dictates that the action of a play should exist in a single physical location. The unity of time determines that the action should occur over the course of a limited timescale. Finally, the unity of action decrees that a play should be defined by a series of principal actions.

The unity of place has been imitated by football analysts in their use of xG shot maps. Figure 3.2 shows one such map, that of Harry Kane, which Sky Sports put out across broadcasts and online media. The gradient of each dot represents each effort's post-shot xG – a metric that we'll study later on. The stars show the shots that ended up in the back of the net, while the circles are efforts which were either saved or missed the target. These maps are sometimes comprised entirely of dots, the sizes of which represent the xG value. A shot

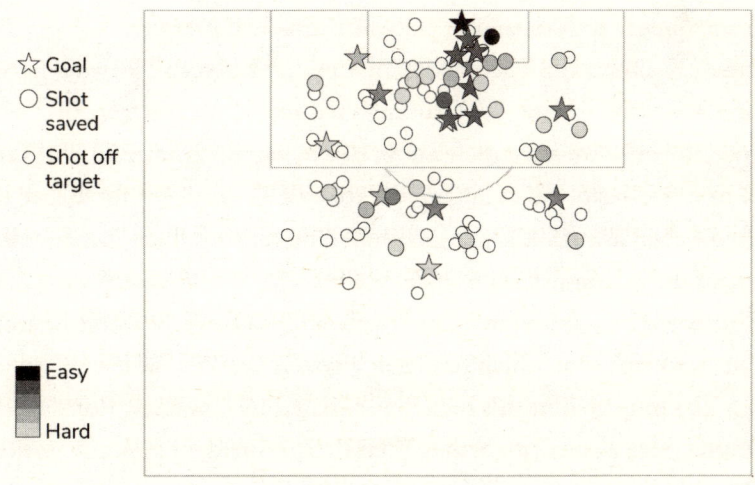

Figure 3.2: Harry Kane Shot Map, Premier League 2020/21

with a high xG value will mean a larger circle, while low xG efforts will be represented by smaller dots. Shot maps have become popular because of their intuitiveness – an xG map clearly shows the quantity, quality, and locations of each team's efforts on goal.

Shot maps offer a great insight into the chances that a player is accumulating. Harry Kane clearly generated a large number of chances in and around the six-yard box during the 2020/21 campaign. At the point in the season when the graphic was put out, early April, he'd also scored four long-range efforts from outside the box. His chances predominantly came from the left-hand side of the penalty area, perhaps an indication of his right-footedness: he prefers to shoot when he can open his body up and curl it towards the far post rather than hitting it across the goalkeeper from the right-hand side of the box. Clubs can use these graphics to identify the areas that opponents like to operate in, or to assess a potential signing's ability to generate scoring opportunities.

Sky Sports' broadcasts have emulated Aristotle's second unity, the unity of time, by featuring analysis of xG timelines. These graphics show the development of each team's cumulative Expected Goals throughout a game. A football match does not consist of a random cluster of attacks from either side. Anyone who has watched the sport will recognise that teams will often exert spells of dominance over their opposition. The tide of a game ebbs and flows. Commentators remark that 'it is important to score while you are on top', because soon enough it will be your opponent's turn to enjoy a period of sustained pressure.

Figure 3.3 shows the xG timeline of a match between Norwich City and Manchester United in late 2021, as analysed on *Monday Night Football*. On the horizontal axis, we have the minute of the match (from 0 to 95). On the vertical axis, we have the cumulative Expected Goals total for either team. Each time either side takes a shot, the line representing them increases by the xG value of the attempt. For instance, the first shot of the match was taken by United in the 11th minute and was worth 0.03(xG). Thus, their line moves above Norwich's. A shot which results in a goal is signified to have a dot at the top of it. The only goal in this match was scored by Cristiano Ronaldo in the 75th minute, from a penalty worth 0.77(xG). This incident is clearly noticeable in the graphic.

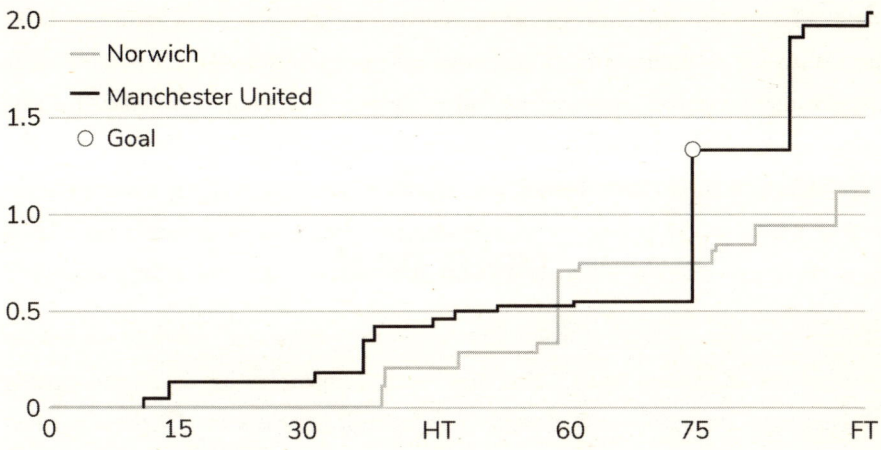

Figure 3.3: xG Timeline Norwich 0-1 Manchester United, 11 December, 2021

The xG timeline gives analysts a unique insight into the spells of pressure which each side endured during the match. Figure 3.3 gives the impression of a cagey game between Norwich and United. Both sides struggled to create clear-cut chances in the first 35 minutes or so. Each team had a half chance to open the scoring just before half time. Norwich missed a good goalscoring opportunity just before the hour mark. With the xG scoreline reading roughly 0.7-0.5 with 15 minutes left, United scored a penalty. Norwich carved out a few low-value opportunities in the remaining time, but struggled to find a winner. Clearly, there were periods when either team was on top, and also spells when the game was more open and spells when the game was more closed up. Expected Goals timelines can tell the tale of any match without the analyst having to watch any of the action.

The final of Aristotle's dramatic unities is that of action – a performance should be made up of a series of consecutive events. The Bundesliga's pioneering use of xG in broadcasts has made this unity a reality. As well as showing Expected Goals data for each team after every match, they've mimicked Aristotle's unity of action by showing the individual xG of goals as and when they're scored. Suppose Joshua Kimmich scores for Bayern Munich, a graphic will pop up in the top-left of the screen showing the xG of his effort. The great advantage of this form of graphic is the precision with which it presents the data. We are able to glean the exact amount of xG that each team or player accumulated in pretty much the exact moment the chance is created. The Bundesliga have also launched explainer videos to educate their audience and have even attempted to combat the confusing term 'Expected Goals' by rebranding it as 'Goal Probability.'[6]

GAFFERS AND XG

A foxier, more analytical tone has certainly permeated football broadcast journalism in recent years. This has been reinforced by managers openly discussing xG in post-match interviews and public appearances. Chelsea manager Thomas Tuchel sat down to talk with the media after his team's defeat to Arsenal in April 2022. The Blues had conceded 3 goals from just 1.12(xG), before a late penalty provided Arsenal with a fourth goal. The result came off the back of a damaging Champions League quarter-final defeat to Real Madrid. Those two legs were preceded by a 4-1 home defeat to Brentford. Chelsea's form was all over the place and the fans were unhappy.

Speaking after the Arsenal match, Tuchel said, 'It was a kind of freak result. But it feels like a pattern because we had the Real Madrid and Brentford games.' That trio of back-to-back home defeats meant Chelsea, who had kept six clean sheets in nine games prior, had since let in 13 goals in six matches. 'We've conceded 11 goals in three home matches,' said Tuchel. Unprompted, he moved on to the xG stats. Chelsea had conceded 4 goals from 2.48(xG) against Brentford, before Karim Benzema scored 4 goals from 1.85(xG) against them over the two legs of their Champions League clash. At that point in the season, Chelsea should have let in roughly eight goals since the March internationals according to xG. In reality, they had let in 13. The Arsenal result had rubbed yet more salt into Chelsea's gaping xG wounds.

When Tuchel was pressed on his own views of xG, he gave an unusually in-depth and academic response for a Premier League manager. 'We've known about it for a long time and now it is out there in public, which I think is good

---

[6] 'The Goal Probability Philosophy' doesn't have as nice a ring to it.

because it gives you a more realistic view on your performance,' he said. 'You can lose games in football by being unlucky and you can win games with luck and the result very often does not reflect what happened on the pitch. So it gives you a clear view: how many chances you allow, how many shots of which quality you allow. It's good to have that figure.' The analysts in the press room were delighted.

'Like with every number,' he continued, 'the more you look into it, you find your benchmarks. So we know over a season or half a season what the level of Expected Goals is if you play in a certain structure and if this is suddenly higher we ask ourselves, "what's happening within the structure?" If we concede double the amount we are tempted to say we're in a very unlucky streak at the moment because obviously the quality that we give away is not enough to concede so much and still we concede. That's hard to take.'

The story of Thomas Tuchel's xG education began almost a decade earlier. At the end of the 2013/14 season, having just guided Mainz to a seventh-place finish in the German Bundesliga, Tuchel decided he needed a break. Despite securing Europa League football for Mainz, he struggled to see how the club could reinvent itself. Tuchel was determined to expand his football knowledge at this point in his career, so he took some time off to learn about more innovative coaching techniques. Tuchel's quest for inspiration led him directly into the path of Matthew Benham.

Brentford were flying high at that point. Benham's team had recovered from Marcello Trotta's 2013 penalty miss to gain promotion to the Championship in 2014 for the first time in two decades. By the turn of the year, the Bees were pushing for a remarkable back-to-back promotion. Tuchel became aware of Benham's work and a meeting was arranged through a mutual contact. They were first introduced at a hotel in Hamburg, where their football philosophies instantly clicked. Tuchel was fascinated by how Benham's data consultancy, Smartodds, used complex algorithms to analyse football matches.

In February 2015, Tuchel came to London for a couple of days. Brentford had controversially decided to part company with manager Mark Warburton (which they eventually did that summer) in order to fully implement their data-driven philosophy. Although it was not a formal interview, Benham launched a charm offensive with the secret ambition of recruiting Tuchel as Brentford head coach. Tuchel was given a tour of the Smartodds headquarters alongside Arno Michels and Rainer Schrey, who had been his assistant and fitness coach respectively at Mainz. The group were granted an exclusive breakdown of Smartodds' intricate football analysis operation.

What was once a set of niche and widely mocked metrics has now become a go-to tool for foxy top-tier managers and players.

Tuchel was particularly intrigued by how they evaluated the likelihood of a team winning a match in real-time, using a system founded on xG data. Tuchel also shared his own nuggets of wisdom. He told Smartodds staff members working on player analytics that he was a big admirer of a teenage striker at Stuttgart – Timo Werner. 'He absolutely loved him and went on forever about his potential and skillset,' a former Smartodds employee recalls.

Tuchel was also intrigued by Smartodds' assessment of Borussia Dortmund's performances so far that season. Jürgen Klopp's team had gone into the Bundesliga's winter break in the relegation zone, but Smartodds' data showed that they had been extremely unlucky. Tuchel ended up taking over from Klopp at Signal Iduna Park a few months later and kept in loose contact with some of the analysts he had met at Benham's offices, calling on them to provide qualitative analysis and in-match data on an informal basis during his stint at the club.

Tuchel's xG education in the Smartodds offices clearly made an impression, given his comments made several years later following his Chelsea team's defeat to Arsenal. Tuchel's interview came the day after Ralph Hasenhüttl had used Expected Goals to defend Southampton's performance in their defeat at Burnley. The list of managers who have used xG to justify their own team's results over the last few years continues to grow: Arsène Wenger, Frank Lampard, Mikel Arteta, Gareth Southgate, Dean Smith, Thomas Frank, and Graham Potter, among countless others. The last three on that list are unsurprising inclusions, having all been schooled in the Brentford and Brighton schools of xG analysis.

Players have also started publicly mentioning Expected Goals. James Maddison has told interviewers of his desire to increase his xG figures. After scoring for Leicester City against Chelsea, Maddison spoke of how his manager and a member of the analytics department had spoken about where goals come from based on xG. 'Me, the gaffer, and Jack the analyst – Jack will be buzzing that I've name-dropped him – sat down and looked where I could get more goals,' he told the assembled media. Meanwhile, Jack Grealish once noted that he had the most Expected Assists in the league. The England star famously told an interviewer he didn't know what an 'encyclopaedia' was, but has no trouble rattling off xA figures. What was once a set of niche and widely mocked metrics has now become a go-to tool for foxy top-tier managers and players.

The science of winning football matches is what pundits and the media are fundamentally interested in. Who are the best teams and players? Why are they the best? How can others get as good as them? The media are getting better at answering these questions as the tone becomes increasingly analytical. The fundamental difference between a fox and a hedgehog, and perhaps the ultimate difference between a good analyst and a bad analyst, is that the former sets out to prove themselves wrong. Foxes are always stress-testing their systems and changing their opinions; meanwhile hedgehogs are usually one-sided in their argument. The increasing number of foxes appearing in the media won't necessarily mean the end for the Roy Keanes of this world. Quite the opposite. As the tone gets more analytical, the importance of light-heartedness will also grow. The utopian ideal is to blend analysis and insight with passion and debate. A world where foxes and hedgehogs live alongside each other in harmony.

CHAPTER SUMMARY

- Prediction-makers can be categorised as either hedgehogs or foxes depending on their open-mindedness and ability to incorporate new information into their models.
- The nature of football punditry means pundits often display hedgehog tendencies, although sports media is becoming foxier.

- Managers and players have started talking about xG more openly, which has helped analytics reach a wider audience.
- Adopting a fox-like mentality is crucial to unravelling the science of winning football matches.
- A blend of serious analysis aligned with warm and lively debate is the best way for the popularity of xG to continue to grow.

4

THE DATA CONSULTANCY
THE BRAINS BEHIND THE ANALYTICAL REVOLUTION

'War is 90 per cent information'

Napoleon Bonaparte

You arrive at the Smartodds office at 11:30 am, as agreed. Upon entrance you're greeted by a small reception area. To the left you spy a glass-walled meeting room housing an oval table and a whiteboard, on which several maths equations have been crudely scribbled with a marker pen. Next to the meeting room stretches an unextraordinary corridor. You don't know it yet, but at the end of this hallway is a larger room, an office space in its own right, where Brentford's analytics department carry out perhaps the most profound and insightful football analysis anywhere in the world.

You're greeted by an employee of the company. 'Very nice to meet you,' he says. 'Please follow me.'

He leads you down the stairs on the other side of the reception area and on to a vast trading floor. You walk through a corridor of desks, each one housing at least four large computer monitors. The men sitting here are fully dialled into their screens. A woolly mammoth could walk in through the front door and go unnoticed. These are Smartodds' clients, professional gamblers who pay to access the company's broad range of tools and data services. The setup is what you imagine the trading floor of an investment bank might look like, albeit with khaki shorts and t-shirts replacing the Armani suits. This collection of desks precedes a separate group of desks at the end of the room, with less impressive tech and only two monitors per station. These worktops are where the data collectors sit, logging stats which feed into the company's advanced algorithms. Framed football jerseys and other items of memorabilia line the bare brick walls. It's approaching midday, but hardly anyone is positioned at these worktops. It's lunchtime on a weekday, so not many football matches are on and in need of analysing.

The man stops at one of the desks. 'Please have a seat.'

You applied for a job at Smartodds on a whim after reading about the vacancy online. The job specification offered a vague description of the role: you'd be contracted as a data collector for a company that specialises in providing in-depth quantitative and qualitative research and analysis in football. Whatever that means.

The process to get you this interview was fairly simple. The company sent you a 31-page document entitled the 'Smartodds Watcher Handbook'. The front cover showed a picture of an anonymous footballer lying down on his back looking up at the sky in dismay, presumably having just missed a big goalscoring chance. Underneath this image was written in red writing, 'Confidential – Property of Smartodds.' The document introduced itself by saying it provided 'a comprehensive guide to the watching process and outlines what Smartodds expects for any person who provides watching services for the company.' It was, in essence, a manual for how to assess the danger of goalscoring opportunities in football matches.

The first page of the document gave a brief history of the Watcher Operations team. The team was formed in 2006 to give real-time match analysis to its clients. They offered a detailed reflection of how games were developing, primarily through providing live data such as shots on target, shots off target, corners, and possession percentages. 'However, such statistics are in many respects restrictive and ambiguous,' the manual explained. 'For example, shots off the post or glaring misses from three yards are classified as shots off target – the same category as a harmless stray shot from 30 yards. Consequently, a new process of rating the danger of attacks was developed with the Watcher Operations Team.'

The method of rating attacks was defined by five core classifications:

- **Delivery**: Any **meaningful attack** should be logged as a delivery. Examples include any time the ball is dribbled into the opposition box but no shot takes place, shots from outside the box which fly harmlessly over the bar, or efforts at goal which are easily blocked by defenders. Attacking set pieces also count as deliveries, including all corners, long throw-ins, and free kicks inside the opponents' half, regardless of whether or not these situations were under- or over-hit.

- **Half Chance**: One step up from Delivery. There was an opportunity to score but **nothing particularly dangerous**. Examples include a shot from the edge of the box which would have required a keeper error to go in, a header from 12 yards which an attacker should be able to direct goalwards but there are other factors going against him, or any free kick from around the edge of the box.

- **Chance**: One step up from Half Chance. There was a **decent opportunity to score**. Examples include if the goalkeeper makes a good save (either at full stretch or fingertips), a shot from 18 yards which went narrowly wide but would have gone in if the other side of the post, the attacker was going to shoot from a dangerous area but the defender makes a last-ditch tackle, or a big goalmouth scramble where the ball could have gone anywhere but no one has real control of the ball.

- **Oooh**: One step up from a chance. An **extremely near miss**, where there seemed to be **at least a 50 per cent chance** of a goal being scored. Examples include penalty misses, one-on-ones, an awkward last-gasp clearance off the line, a cross flashed across goal which narrowly misses the boot of a striker from close range, or any attempt which hits the post or crossbar. This rating gets its name from the sound that can be heard inside a stadium when a team misses a glorious chance.

- **Goal**: Self-explanatory.

The guide also instructed you to award ratings based on controversial refereeing decisions. If an attacker is tackled clumsily in the box but no penalty is given, that merits the awarding of a Half Chance to the attacking team. If a team score a goal but VAR rules it out for a marginal offside, a Chance should be awarded. If the referee fails to play advantage when the team are about to be through one-on-one, an Oooh should be logged. In this sense, the way Smartodds assess chance creation is superior to standard xG models, which would log these situations as 0.00(xG) as no shot has actually taken place.

The Watcher Handbook is keen to stress that only one rating can be provided from an individual attack. Suppose two shots happen in quick succession that each merit the awarding of a Chance. Rather than logging two separate chances, you should upgrade the whole sequence to one Oooh. The Expected Goals method deals with sequences which feature more than one shot in a similar way. Consider a Manchester City penalty which took place against Watford late in the 2019/20 season. Raheem Sterling's spot-kick was saved by Ben Foster, but the ball rebounded to Sterling who tapped in from close range. Penalties are always worth 0.77(xG) and the rebounded shot was worth 0.90(xG). As such, an analyst might be tempted to award Manchester City 1.67(xG) from this sequence, but this would be wrong because it's impossible to score more than one goal from a single attack. So how does xG deal with this problem?

The key is working out the probability that the attack *doesn't* result in a goal. In other words, how likely is it that Manchester City don't score either of these chances? The maths behind this particular situation is as follows:

$$(1 - 0.77) \times (1 - 0.90) = n$$

$$0.23 \times 0.10 = 0.02$$

So there's a 2 per cent probability that Manchester City don't score either chance. Therefore:

$$1 - 0.02 = 0.98(xG)$$

So Manchester City should be awarded 0.98(xG) for this attack.

Smartodds' data collection system essentially measures the danger of each attack which takes place, and can be considered a variant of Expected Goals. In many ways, it's more powerful than standard xG, which only measures shots that actually take place. A situation where a striker rounds the goalkeeper but runs out of room and ends up dribbling the ball out of play won't be registered in xG. Neither will a ball flashed dangerously across the face of goal that narrowly eludes an on-rushing forward. Smartodds' watchers are trained to log such situations as Chances or Ooohs, meaning these dangerous attacks are accounted for in the dataset.

Along with the handbook, the company also provided 100 video clips of goalscoring opportunities and tasked you with assigning a rating to each one based on the danger scale outlined above. You'd worked your way through the clips over the course of the last week or so. The short videos came in sets of five and after each set, Smartodds' bespoke software would provide iterative feedback on your performance. If you had correctly assessed how dangerous the scoring opportunity was, you'd get a green tick next to that clip. If you'd made a mistake, the computer would tell you what you'd done wrong:

Sorry, you incorrectly marked this 'Chance' as an 'Oooh.' Notice how the player's heavy touch narrows the angle and reduces the danger of the attack.

That's correct, this attack was a 'Delivery' that carried little threat.

Unlucky, that opportunity is actually a 'Half Chance.' Look at the number of defenders between the striker and the goal.

Clearly your assessment of these 100 chances had been accurate enough to advance you to the final stage in the process. You take a seat at the desk as the 'interviewer' turns on both of your screens. On the left-hand monitor he loads up a full replay of a match between Heidenheim and Kaiserslautern in the second tier of German football. On the right-hand monitor, he loads up the company's internal data-collection system. You're tasked with watching the entire game and rating every attack that happens according to the danger scale outlined in the handbook. Afterwards, your performance will be assessed and, if you're accurate enough in separating your 'Chances' from your 'Ooohs', you'll be offered a contract with the company. This is certainly different from any job interview you've ever done before.

You complete your task, clicking the 'Delivery' button every time a weak attack takes place, the 'Half Chance' button every time a long-shot ends up comfortably wide, and so on. The match finishes 3-1 to Heidenheim, but the system at the end of the game shows that Kaiserslautern had actually created more Half Chances, Chances and Ooohs than their opposition. A few weeks later you receive an email. You are now officially a Watcher at Smartodds.

WORLD-CLASS PREDICTION-MAKING

Over the next year or so, you watch hundreds of matches for the company. You're paid £20 per game, with an additional £20 if the match goes to extra time. You watch up to four matches per day, taking small breaks in between games to re-energise yourself with a Mars bar from the vending machine before embarking on another 90-minute stint. Some of the games are recorded matches which took place over the course of the previous week. The stats for these matches presumably go into the large bank of data that the company stores. Other matches are live. These games are particularly interesting to work on because the data feed is sent in real-time to the company's clients, football traders sitting just across the room from you. They are also watching the game live and following the Deliveries, Ooohs and so on which you're logging. This data helps inform their opinions and predictions. Alongside your logging of the company's bespoke statistics, you are tasked with providing a brief update on the flow of the match every 15 minutes. You're asked to note which team is on top, how cagey the game has been, any tactical insights you can give, and other information that might provide context to the raw stats.

Most of the games you watch are from obscure leagues. The company houses satellites that pick up the feeds from a number of different countries worldwide. One minute you're watching Slovan Liberec take on Slavia Prague in the Czech first division, the next you're logging data as Farense play Atlético CP in the Portuguese second tier. You watch a fair amount of lower-league football from England as well and occasionally even getting assigned to a low-level Premier League clash.

The unique proprietary data which Smartodds collects is the foundation of their success. The business has built up a bank of chance-quality data which tells them the danger rating of every attack which occurs in professional football. From this, they can build a picture of how good every team is at creating and preventing goalscoring opportunities. They have an accurate gauge of the attacking and defensive ability of every team. Smartodds were one of the first organisations to uncover the xG treasure chest, and have been living off the riches ever since.

The company's models do also incorporate factors besides the ability of teams to create dangerous opportunities. Smartodds found that a team's chances of winning away from home were slightly raised if they had a shorter journey to the game. This means local derbies are marginally easier to win than games on the other side of the country. A network of international informants also provide 'soft' information which is factored into Smartodds' predictions. Dressing-room strife or an injury to a key player could shift the odds of a team's success.

The brightest guys in football don't tend to originate from football. They've moved over from other industries – banking, technology, consulting and so on. One industry stands alone from others in terms of how many football innovators it's produced: the gambling industry. These people aren't your amateur gamblers, the blokes down the pub placing a bet on Saturday afternoon to receive their weekly dopamine hit. These are the professionals. The ones who are using sophisticated algorithms to guide their bets, who approach football betting as a trader would approach the stock market.

Matthew Benham, the owner of Brentford and Smartodds, once said, 'I never bet for fun or the thrill. We are all about calculating probabilities with the help of mathematical models.'

Benham studied physics and worked as a derivatives trader in the City. It was there he learned the art of exploiting inefficient markets, a skill he would later bring to sports betting. Seeking out undervalued propositions is the key to succeeding in any market, whether that be of the financial, betting or player transfer variety. Just down the road from Smartodds' Kentish Town office resides Brighton & Hove Albion owner Tony Bloom's Starlizard, a company with a near-identical structure and business model to Smartodds. Both of these men live and die by the guidelines outlined in this book. They fully understand the science behind winning football matches and are incredibly skilled forecasters of footballing outcomes. Their data consultancies are the brains behind the decisions being made at their clubs, and are the primary factors why both Brentford and Brighton have been able to haul their way up the Football League ladder. The one rule in the Smartodds offices is that you're not allowed to celebrate a goal. There could be someone sitting at an adjacent desk who has lost thousands because of that goal. But the thing is, none of these bettors are likely to cheer a goal anyway, nor would they be likely to commiserate one. Their approach to betting is like trading. They work on the percentages. They understand luck and probability, which arms them with a thick skin to the whims of chance and protects them from emotional misguidance. But above everything else, they know the science of what wins football matches, and thus know that the odds are in their favour. To get worked up over a single goal is as unproductive as a casino manager getting worked up over a single hand of black jack.

CHAPTER SUMMARY

- Calculating the xG probability for a series of back-to-back efforts on goal, rather than just being restricted to a single effort, is a speciality of Smartodds' analysis and gives a more rounded insight into a goalscoring opportunity. Smartodds' ability to provide this analysis is a big difference between their system and traditional xG theory.

- Smartodds' set-up is reminiscent of a trading floor or an investment bank, reiterating the data-first nature of the company's approach.
- Goals are not celebrated at the Smartodds' office, reinforcing and emphasising the sense that science lies behind a game of football, not emotion.
- Smartodds' clinical study of the game helped the company identify that the distance a team travels for an away match can have an adverse effect on performance.

5

THE BOARDROOM

A TOP-DOWN APPROACH TO ANALYTICS

'Do not follow where the path may lead. Go instead where there is no path and leave a trail'

Ralph Waldo Emerson, 19th century American writer

Throughout history, a handful of significant individuals have possessed an innate sense of purpose and direction, knowing exactly what they wanted to be when they grew up from a young age. From the artistic genius of Pablo Picasso to the scientific curiosity of Marie Curie, these figures pursued their passions from an early stage with unwavering determination and dedication. John Henry falls into the same bracket. Henry grew up playing APBA Baseball, a dice game where the actual performances of major leaguers are rendered on to cards representing each player. Stan Musial, hitter for the St. Louis Cardinals and Henry's favourite baseballer, was just as likely to hit a home run on Henry's bedroom floor as he was at Sportsman's Park. APBA Baseball requires players to use maths and statistics to successfully run a baseball team. These days John Henry enjoys a real-life version of the same game, only on a much bigger scale.

Henry first learned how to exploit market inefficiencies by trading corn and soybean futures in his mid-20s, thus beginning a life-long fascination with statistical trends in markets. Henry developed a mechanical trend-following method for managing a futures trading account. When this methodology proved successful, he founded John W. Henry & Company in 1981. The firm relied on Henry's system of making automated, non-emotional trading decisions in response to systematic determinations of reversals in each market's direction. In simple terms, his model found success by removing the damaging impacts of human emotion. Before long, Henry's net worth ran into the tens of millions.

After acquiring his fortune, he made various ventures into the business of baseball during the 1990s. His headline acquisition came in 2002 when he bought the Boston Red Sox, a team that had been struggling for success over the course of several decades. John Henry was about to shake everything up. He sought to bring the same rigorous, unemotional analysis that had brought him success in the financial markets to the running of a baseball club. 'People in both fields operate with biases and beliefs,' Henry later wrote in a letter to ESPN's Rob Neyer. 'To the extent you can eliminate them both and replace them with data, you gain a clear advantage.'

One of John Henry's first tasks as owner was to fill the vacant general manager position. The name on the top of his shortlist was Billy Beane, who at that time was general manager of the Oakland Athletics. Although Beane had not yet achieved fame through Michael Lewis' *Moneyball* – which would reach publication and world-wide acclaim the following year – Henry was aware of his data-driven and market-exploiting ways of working because Beane had tried to sign Kevin Youkilis, affectionately known as 'the Greek god of walks', from Henry's Red Sox the previous season. Youkilis' name was the first of any college player churned out of the Oakland Athletics computer before the 2001 draft, but had been ignored by the team's scouting department because of his weight and appearance.

Youkilis played as if he was trying to break the world record for walking and wearing out the arms of opposing pitchers. Billy Beane and his analysts realised these skills were drastically undervalued assets in the market for baseball players and could be acquired on the cheap. Beane and his A's team tried to trade Youkilis from the Red Sox mid-season but was outmanoeuvred by Theo Epstein, a 28-year-old Yale graduate who was the Red Sox assistant general manager. Epstein idolised Beane. The two spoke occasionally and Beane was aware that Epstein was one of the few people shrewd enough to appreciate Youkilis' actual value. What Beane didn't know at that point was that Epstein had recently acquired greater power. The new Red Sox owner, John Henry, had begun listening to everything the data-savvy Epstein had to say. Three months earlier, the Oakland A's would have been able to sign Youkilis. As it so happened, Epstein blocked the move and established Youkilis as the poster boy for the Boston Red Sox farm system.

When John Henry offered Billy Beane the general manager role at the end of that season, Beane declined the record-breaking salary of $12.5 million in order to remain with the Oakland A's. For him, the validation that someone in the sport of baseball had recognised the achievements of his team was more important than the money on offer. Who did Henry turn to instead? Theo Epstein, who became the youngest general manager in the history of the sport. He was an unproven, relatively inexperienced candidate, but would become one of the most influential executives in all of baseball. 'Theo had a lot of ideas about how to strengthen the team competitively,' said Tom Werner, the Red Sox chairman. 'He is a very smart analyst of the game who believes strongly in analytics'. Epstein moved the general manager's office closer to the rest of the baseball operations staff and surrounded himself with similarly young and ambitious executives. He launched a total philosophy shift towards exploration, learning, problem-solving and innovation. The mindset originated from the top, with John Henry, and worked its way down through the newly appointed management team.

On the performance analysis side, Theo Epstein's Boston Red Sox signalled their intentions early on by hiring Bill James, the founder of the 'sabermetrics' movement. James was the leading light of baseball's data revolution thanks to his *Baseball Abstract* book series. At that time, James was considering offers from other organisations for front-office opportunities. His worry was that such a position would leave him 'hanging from the bottom of the airplane' without an opportunity to get his hands on the controls. Many of his friends within the analytically minded community had ended up in such positions: working at clubs, but largely ignored and without the ability to make any real impact. The Red Sox convinced him this would not be the case. Epstein's team demonstrated a genuine interest in how James's work could give them a competitive advantage. John Henry, a long-time reader of the *Baseball Abstract* series and subscriber to the Jamesian approach to baseball, was involved in the interview process. The Red Sox owner had long since used Jamesian statistics to 'clean up' in his fantasy baseball leagues. No wonder Bill James got the job.

The triad of John Henry, Theo Epstein and James transformed the Red Sox. After a couple of years, the club had established an analytics department which was light years ahead of the nearest competition. John Henry had recognised the opportunity to gain an edge, brought in experts from a broad range of fields and empowered them in a way never seen before in baseball. They adopted a similar approach to the Oakland A's, but with the advantage of greater revenues and resources. The team surmounted a seemingly insurmountable obstacle in 2004 by winning their first championship since 1914, beating John Henry's childhood team, the St. Louis Cardinals, in the World Series. They won a second championship in 2007 with a team so dominant that it proved beyond any reasonable doubt that the Red Sox had found a winning formula. And that got the ownership dreaming bigger.

The 2007 success fuelled Henry's desire to bring this analytical approach to other sports and remits, whether that be NASCAR, NHL or even the real estate industry. Two years after purchasing the team, the Red Sox ownership created Fenway Sports Management (which later became Fenway Sports Group) and began exploring other entrepreneurial ventures. By the summer of 2010, FSM was meeting with NBA league officials about opportunities to expand their portfolio into the sport of basketball. Nothing materialised at the time, but another intriguing option unfolded. Joe Januszewski, who worked in the Red Sox corporate partnerships department, was an ardent fan of Liverpool Football Club, a fact he made well-known around the office. The club was going through a difficult period and Januszewski often joked about his team needing saving by someone with deep pockets and a strong plan.

On a quiet Tuesday night in August 2010, Januszewski was sitting in front of the television nursing his infant son when his phone rang. It was Larry Lucchino, President of the Red Sox. Lucchino was sitting with John Henry in an executive lounge at the Rogers Centre in Toronto, watching the Red Sox play the Blue Jays.

'Be brief, but tell us what's going on at Liverpool and why it's a good opportunity,' Lucchino said.

The din of the baseball game drowned out Januszewski's elevator pitch, so he turned to the medium of email. He put 'Save my club!' as the subject line and clipped an article from the Sports Business Journal that said Liverpool were headed towards bankruptcy. Two months later, Fenway Sports Management owned Liverpool FC.

The club John Henry inherited was a failing enterprise. Liverpool had finished the prior season in seventh place and were paying interest payments on loans of £340,000 per week. The question was whether Henry, who had successfully used analytics to conquer both the financial markets and baseball, could use these tools to turn Liverpool's fortunes around. Would Henry's methods bring him success in the world's most complex, dynamic sport?

At first, the answer appeared to be a resounding 'no'. In Liverpool's first six seasons under FSG's ownership following the purchase in October 2010, they finished above sixth place only once (in 2013/14, the year of club captain Steven Gerrard's infamous slip against Chelsea). Their only Champions League campaign over this period saw them knocked out in the group stage. The battle at the top of the Premier League was becoming more intense than ever and, from an external viewpoint, it seemed like FSG were out of their depths.

Fenway Sports Group couldn't outspend the sheikhs and oligarchs who were pumping money into rival teams. They had to be smart. But their analytical philosophy was struggling to have an impact due to resistance from within. The relationship between the front office that FSG had installed and Brendan Rodgers, Liverpool's manager at the time, was incredibly strained. Rodgers was reluctant to give up power to the club's transfer committee. He pushed through signings like Mario Balotelli and Christian Benteke, who ultimately flopped. At the same time, he ignored the 'data signings' that had been recommended to him by the analytics department. Rodgers shunned the highly-rated duo of Iago Aspas and Luis Alberto, signed in the summer of 2013 based on the algorithms of Michael Edwards, the Director of Technical Performance. Aspas was handed just five League starts, while Alberto was given only a handful of substitute appearances in the league. Roberto Firmino was purchased for £29 million, also based on Edwards' statistical model, but Rodgers played Firmino on the right-wing despite the analytics department's assertions that he'd been signed to play down the middle.

The clash of Brendan Rodgers' traditional style of management and the insight provided by the data department meant the club was fatally misaligned. The club's fortunes suffered as a result and, unsurprisingly, the analytics department took the blame. Many pundits believed that Liverpool's reliance on numbers was undermining the 'football men' who should have been making the decisions. *The Independent* wrote that the club's biggest problem was their 'deep attachment to the theory that statistics and analytics can provide the answers.'

A poor start to the 2015/16 Premier League season was the final straw. Fenway Sports Group decided to go all-in with their analytical approach. They sacked the traditionally minded Brendan Rodgers and went out in search of a new, progressive manager who would work within the club's data-driven framework. They wanted a coach who could lead the team and guide the culture, but who was also fully committed to the organisation's radical restructuring and wholesale adoption of analytics.

INSIDE LIVERPOOL'S DATA REVOLUTION

Jürgen Klopp was still in his first month as Liverpool manager, in November 2015, when Ian Graham came knocking. The team's Director of Research arrived at Klopp's office door armed with computer printouts folded under his arm. Graham hoped to show Klopp, whom he hadn't yet met, how Liverpool could benefit from his analysis. Then he wanted to convince his German boss to actually use it. Graham spread the papers out on the table in front of Klopp, which showed information on a game between Borussia Dortmund, who Klopp was in charge of at the time, and Mainz from the previous season. The Klopp-coached Dortmund team had dominated the smaller club but had ended up on the wrong side of a 2-0 defeat. Before Graham could begin to talk through the printouts, Klopp's face lit up. 'Ah, you saw that game,' he said. 'We killed them. It was crazy. You saw it!'

Ian Graham had not seen the game. But a couple of months earlier, as Liverpool were deciding who should replace Brendan Rodgers, Graham had carried out a statistical analysis on the performance of each of the leading candidates. The Expected Goals data advanced the notion that Klopp's Dortmund had performed a lot better than the league table suggested. Based on xG, they should have scored roughly seven more and conceded about nine fewer goals than they actually did over the season. Dortmund finished seventh in the Bundesliga, but Graham's model showed they were actually the second-best performing team. 'We really thought that Dortmund were unlucky,' Graham said in early 2019. 'It wasn't anything to do with systematic problems with the players and the coach. If we wanted to hire Jurgen as our new manager, we had to understand if there was something wrong with Dortmund. And the statistical analysis said very clearly and very strongly: no'. Even though Klopp's reputation had suffered as a result of his club's final league position, Graham had seen that their disappointing season was an illusion. It had nothing to do with Klopp's ability as a manager; he just happened to have been coaching one of the unluckiest teams in recent history.[7]

In that game against Mainz, Ian Graham's printouts demonstrated that Dortmund took 19 shots compared to their opponent's 10. They worked the ball into Mainz's penalty area on an impressive 36 occasions; their opposition

---

[7] Ian Graham's analysis found that Klopp's Dortmund were the second-unluckiest team over the previous 10 seasons of the Bundesliga.

managed only 17. Dortmund missed a penalty in the 70th minute, only to score an own goal four minutes later. Overall, Dortmund created 2.13(xG) to Mainz's 0.93(xG). Based on these xG figures, there was only a 10 per cent chance that Dortmund would lose this game. They dominated by every conceivable measure, except the final score.

Ian Graham then pointed Klopp towards a different set of printouts, detailing a game played a month later between Dortmund and Hannover. The data was weighted even more in favour of Klopp's team: 18 shots to seven, 55 balls into the box compared to 13, 1.69(xG) to 0.72(xG). Once again, Klopp bemoaned his team's misfortune in that game and voiced his surprise that Graham had watched the match.

In reality, Graham had not seen the Hannover game, either. In fact, he hadn't seen any of Dortmund's games that season, neither live nor on video replay. He didn't need to. In order to make his assessment on whether or not Klopp was suitable for the Liverpool managerial role, he'd only needed to consult the xG data.

Klopp didn't utilise analytics during his time at Dortmund. Like most managers at that time, he was all-consumed by coaching his team out on the training pitch and the field of play. But by the time Graham left his office that morning in 2015, Klopp was an xG believer. Despite the fact Graham hadn't seen any of Dortmund's matches that season, he had convinced Klopp of his appreciation of the bad luck that had befallen his team. Klopp later learned that without Graham's analysis of that campaign, he never would have been hired. 'The department there in the back of the building?' he later said, referring to Graham and his colleagues. 'They're the reason I'm here.'

Dr Ian Graham is a lifelong Liverpool fan. He grew up just outside Cardiff, nearly 200 miles from Anfield, but his childhood aligned with Liverpool's era of total dominance throughout the 1970s and 1980s. He idolised their striker, Ian Rush, who just so happened to be Welsh. Graham realised he didn't want to be a scientist halfway through a two-year post-doctorate at Cambridge University. Polymer physics, his area of study, had seen most of its breakthroughs happen in decades gone by and there wasn't a lot of ground left to break. Graham stumbled across a job opening for an analytics start-up that was hoping to consult for football clubs. He landed the job, before promptly buying a copy of Michael Lewis' *Moneyball*.

For four years, from 2008 to 2012, Graham consulted at Tottenham. Or at least, he 'consulted' in the loosest meaning of the term. The managers who ran the club over that period had little interest in his consultation. The struggle of football analytics has been one of getting engagement from those who run clubs. Countless stories have been told of analysts finally breaking into the inner sanctum of a team, only to be shunted to a corner and ignored by the head coach. **Modern-day managers can no longer be seen to not have an analytics department**, but they're under no obligation to pay attention to them once they're there. Fenway Sports Group wanted to be different. They wanted to build a team with analytics at its core. They wanted Ian Graham.

When FSG bought Liverpool, they hired Graham to build a replica of the Boston Red Sox's research department. The reaction from the staff already inside Liverpool Football Club was less than favourable. Graham and his team were touted as 'laptop guys' who 'didn't know the game' according to Barry Hunter, who was installed in Liverpool's scouting department. Graham was unfazed. He became completely absorbed in his quest to defy conventional football wisdom and bring success to his childhood team.

Another key cog in the Liverpool backroom staff was Michael Edwards, a man who keeps a notoriously low profile, never speaking to the press and rarely making public appearances, but who has nonetheless been instrumental in Liverpool's recent success. Edwards showed a hint of what was to come when, while playing for Peterborough U18s, he took a keen interest in computing. His teammates teased him for throwing himself into an IT module that the rest of the youth squad considered a waste of time. After leaving the club, Edwards studied business management and informatics at Sheffield University before landing a job as an analyst for Prozone, a budding football data company. The business operated out of a warehouse in Leeds, where its small contingent of part-time processing staff would film matches and then meticulously break them down into datasets that could be used to objectively analyse the performance of teams. At the time, the work done there was pioneering.

The company was looking to place their analysts at professional clubs around the country. Edwards got assigned to Portsmouth, where he was responsible for analysing the first team's performances, presenting the tactics of upcoming opponents, and evaluating transfer targets. But the year was 2003 and football wasn't ready for the data and technology revolution that Prozone were looking to launch. Edwards once received a phone call from Portsmouth manager Harry Redknapp early in his placement at Fratton Park. The gaffer had rung him to complain that he had put a CD-ROM containing player data into the CD player in his car and couldn't fathom why it wasn't playing anything. For younger readers, that's the equivalent of trying to put a USB stick into an iPhone.

The players were equally as inexperienced when it came to analytics, but Edwards quickly struck up a rapport with them. He wasn't afraid to tell even the more illustrious members of the Portsmouth team, which housed the likes of David James, Sean Davis and Peter Crouch, that, statistically speaking, they'd been very disappointing that weekend. If Edwards had an opinion, he'd let you know it. The players started coming to him, asking about any weaknesses they could exploit in upcoming opponents. Such questions were usually reserved for the manager, but Edwards had gained the respect of the team. His input wasn't just worthwhile, it was actively sought out. Edwards even ran a Champions League prediction game with the players. Each week he'd collect score predictions and whoever finished bottom of the table would have to drive a Robin Reliant to training, and add an outlandish modification to the vehicle in time for the next Champions League gameweek.

Harry Redknapp left Portsmouth for Tottenham in October 2008 and recruited Michael Edwards to join him a year later. Edwards' stint in the Spurs analytics department lasted two years, before he was poached by Liverpool and installed by John Henry as the Head of Performance and Analysis. He served an important role in setting up the club's initially much-maligned transfer committee. Over the years he worked his way up the ladder, eventually serving as the club's

Sporting Director. His role involved trying to cater for the medium- to long-term interests of the club. His position was fairly revolutionary in English football at that stage. The fact that Edwards could focus on longer-term strategies meant Klopp could throw himself into the day-to-day coaching of the squad.

From the outset of his tenure, Klopp embraced the analytical work the club was doing behind the scenes. He met semi-regularly with Edwards, Ian Graham, and other members of the analytics department to discuss player recruitment and tactical matters. Klopp became skilled at translating the data and research of the performance analysis department into easily digestible team talks and strategy meetings with the squad. Before each game, Graham and the three analysts who worked under him compiled a packet of information. By the time these insights reached the ears of the players on the training pitches or the changing rooms, the PhD-level equations were long gone. Liverpool's playing staff were only faintly aware that the tactical advice was rooted in doctorate-level mathematics. 'We know someone has spent hours behind closed doors figuring it out,' said midfielder Alex Oxlade-Chamberlain. 'But the manager doesn't hit us with statistics and analytics. He just tells us what to do.'

Klopp's ability to blend traditional coaching methods with pioneering data analysis quickly led to a change in the club's fortunes. Liverpool finished in the top four in each of his first six full seasons at the club, winning their first ever Premier League title in 2020. They reached the Champions League final in three of the five years between 2018 and 2022, lifting the trophy in 2019. Between 10 March, 2019 and 24 February, 2020, Liverpool played 36 league matches. They won 35, drew one and lost none. The Reds accumulated 106 points from a possible 108 over this period. Luck obviously plays a part in forming such an unbelievable record – the best record over a 36-game period in league history, no less – but the performance of the team over this timeframe was a world away from where the club had been before FSG took the reins. Fenway Sports Group had returned Liverpool to the pinnacle of world football.

And they'd done it on a much tighter budget than their competitors.

Ian Graham's role, alongside preparing data packages before matches, was to assist in the recruitment of players. Here, too, the new structure of Liverpool gave them an advantage. Graham built his own database to track the progress of more than 100,000 players from around the world. The power of transfer veto had been taken out of the hands of the manager and placed into the hands of a transfer committee, on which sat the manager, members of the scouting team and members of the analysis department. Instead of relying on one man's knowledge, as is football tradition, Liverpool began making decisions based on the shared wisdom of the collective. Much of this wisdom, of course, is founded on the advanced statistical analysis of Graham and his team.

In 2017, Graham's model identified Mohamed Salah as a transfer target worth looking at. Salah's record in England left a lot to be desired. He scored two goals in 13 games over two seasons at Chelsea, spending lots of time loaned out at other clubs. Graham's data suggested Salah would pair up well with Roberto Firmino, already on the payroll at Liverpool and creating more xG with his passes than almost any other player in his position. Jürgen Klopp preferred Julian Brandt, a youngster who was making waves in the Bundesliga, to Salah. Brandt seemed the obvious choice, but Graham and his model won out in the end. Liverpool paid Roma $42 million for Salah, who went on to score 32 league goals in his first campaign at the club – a record in a 38-game season in the Premier League era. The Egyptian became a symbol of Liverpool's revival under Fenway Sports Group.

Perhaps Ian Graham's most important acquisition came a few years earlier. One of his first assignments under the FSG was to research a forward who was playing at Inter Milan, Philippe Coutinho. Graham's model approved of the young left-winger, who Liverpool ended up signing for $16 million. Coutinho contributed greatly to the club on the pitch over the next few years, but perhaps his greatest contribution came in the windfall he facilitated when Barcelona bought him in 2018. Liverpool sold Coutinho for $170 million, a whopping $154 million more than they'd bought him for five years earlier. The money was reinvested in several new signings, once again recommended from Graham's model. Among them were Alisson Becker, Virgil van Dijk and Fabinho, the core of the team which drove Liverpool to Champions League and Premier League success.

Jürgen Klopp clearly played a huge part in Liverpool's ascension. But his appointment only tells half the story. The team of veiled figures who were installed by John Henry guided the strategy and direction of the team. Behind the scenes, these figures lent on xG data and several other advanced metrics to influence decision-making. This hidden team of number crunchers were happy to let Klopp take the public acclaim. The manager acted as the outward face of the operation, allowing Liverpool's team of analysts to fly under the radar. But these puppet masters were equally as, if not more, responsible for Liverpool's return to prominence as Klopp or any of his players. FSG's restructuring of the Liverpool organisation is what set them apart from the competition and allowed them to compete with the likes of Manchester City, Chelsea and Manchester United, despite possessing far shallower pockets.

Liverpool's approach under FSG is what Isaiah Berlin would describe as 'foxy.' They deploy a transfer committee which takes multiple viewpoints into account. They utilise knowledge from spheres outside of football – their analysis team was headed by Graham, a PhD physicist, and counted in its midst a former junior chess champion, an astrophysicist, and a former CERN employee who helped verify the existence of the subatomic Higgs boson. Even their owner, John Henry, initially achieved success from an algorithm that predicted fluctuations in the soybean market. The same sort of analysis is knit into Liverpool's DNA. The club's ownership have constructed a world-class side while maintaining a competitive wage balance. And they've done it by taking power away from emotional individuals and placing it into the hands of cold, hard data.

CHAPTER SUMMARY

- John Henry's adoption of statistical methods in business, and then baseball, was a reflection of the broader trend of using data analytics in sports and also its growing importance across the commercial and financial world.
- Liverpool FC initially demonstrated the type of traditional mindset that has long been at the forefront of football management, namely that it is people, not data, that make decisions. However, that soon changed when Jürgen Klopp embraced the new-found analytics and grasped its effectiveness and potential.

- Liverpool FC's data analytical team was composed of people from diverse backgrounds. This integration was critical to the success of the club as it brought together expertise from various disciplines, underlining that there is no one 'perfect' collection of analysts; people can add their own strengths into a wider system.
- Data-driven approaches require human 'buy-in' to succeed – neither can work without the other.

6

THE MANAGER'S OFFICE

A NEW BREED OF HEAD COACH

'We re-designed the club based on one question: "what would a football club look like if it had no human eye and ear?"'

Rasmus Ankersen, Southampton (and former Brentford) Director of Football

The Brentford manager, Mark Warburton, was sitting in his office on a cold February morning in 2015 when the story broke.

'*Rayo Vallecano head coach turns down Brentford approach,*' read the headline in the *Mail on Sunday.*

Warburton's side had enjoyed a positive victory the day before and were sitting in the play-off places in the Championship. They were pushing for promotion to the Premier League despite possessing the 21st biggest wage budget in the 24-team division.

David Weir, Brentford's assistant manager, turned to Warburton, 'I think that's a game-changer, gaffer.'

Warburton picked up the phone and called his owner, Matthew Benham.

'What's going on Matt?'

'I can't lie to you, we need to have a chat.'

'Lunch tomorrow?'

They had lunch the next day. It was confirmed that Mark Warburton, the man who had led Brentford to the Championship and was on the brink of pulling off miraculous back-to-back promotions, was to leave the club when his contract expired at the end of the season – regardless of whether the team reached the promised land or not.

When rumours of Warburton's departure first reached the fans, they assumed he'd been poached by a bigger club. As is the way when a manager outperforms expectations at a small club, a big club will eventually come knocking. Brentford's fearless playing style had put Warburton's name on the map. They'd swept aside clubs 10 times their size financially, the likes of Leeds United, Fulham and Nottingham Forest. Their attacking brand of football had earned them and their manager admirers all across the country. Speculation swirled that Norwich City were interested in securing Warburton's services, with Premier League interest also rumoured on the football grapevine.

Then another story broke. *The Times* published a piece alleging that Warburton would be 'sacked' at the end of the season, irrespective of where the already over-performing team finished in the table. (This statement was true, apart from the allegation of him being fired. In reality, Warburton's contract wasn't being renewed. But 'not renewed' doesn't sound as sexy as 'sacked'.) Fans and the media quickly turned on the club. The narrative was no longer that Brentford were the plucky newcomers sticking it to the big boys. They were now the laughing stock. Who do they think they are, casting aside the man who led them from League One obscurity to the Championship play-offs places? *The Guardian* posted an article headed, '*Silly Party candidates bidding for a majority in football's boardrooms*' which compared Benham to a string of recent tyrannical owners to have taken over at Football League clubs.

The Brentford players made their feelings known the evening of *The Times* article's publication. When Andre Gray opened the scoring against Watford, he and his teammates ran straight to the touchline to mob Mark Warburton. The Bees went on to lose that game 2-1, then got beaten 3-0 by Charlton Athletic a few days later. Shortly after the Charlton hammering, the club finally broke its silence on the situation. Brentford released a statement confirming the departures of Warburton and his coaching assistants that summer.

What had been brewing behind the scenes at Brentford before this dramatic story broke serves as the perfect vignette for the struggle that analytics has faced in conquering football. The January transfer window had come around a month earlier. Brentford had been the surprise package that season. Many had tipped them for an immediate relegation back to League One, but they'd defied the odds and were third in the table on Christmas Day 2014. However, things were not looking so rosy according to Matthew Benham's mathematical model. Smartodds had crunched the xG numbers and seen that Brentford were overperforming; their results had been better than their performances should have allowed. The Bees were lucky to be third in the table. In fact, the xG data showed they'd been the 11th-best-performing team and were likely to regress in the second half of the season. Benham determined the best chance of maintaining their promotion push was with new additions to the team, many of whom they'd already identified via their data-driven scouting tools. Moreover, the current squad was small and likely to tire as the brutal 46-match campaign wore on, making new recruits even more of a necessity. According to Benham's analysis, Brentford's chances of promotion would be boosted by more than 10 per cent with a sprinkling of new players in vital positions.

Warburton disagreed. His view was that the tightly-knit dressing room had got the club that far and new arrivals would disrupt the squad's harmony at this crucial stage of the campaign. The dressing room was strong and solid and didn't need changing. As Benham became increasingly determined to empower his data analysts, Warburton became increasingly resistant. It was as if both men were stuck in quicksand; the harder they tried to pull free, the more they found themselves stuck. Warburton would state that you can't use data to measure factors like team unity, combativeness and togetherness. He would claim the team had momentum. Benham's model told him a different story.

The two camps had set their stalls out. On one side, Warburton, the traditional 'football man' who knew he'd have the backing of the football world and plenty of leverage in the form of job offers if he didn't get his way. On the other side, Benham, the analytical 'outsider' who dared to dream of a new model for his club, a world where the manager wasn't all-powerful and where data was harnessed to achieve an edge. The dispute eventually reached a point where Warburton and his team banned Benham's analysts from accessing the training ground. This is perhaps the most literal representation of the struggle that analytics has faced in sport. 'Football men' holding the keys to the palace and refusing entry to science. The culture war had reached its pinnacle. Warburton had locked the door to analytics, both ideologically and physically. If Benham and his ideas were to prevail, he'd have to break the door down.

And that's what he did. He went out in search of a coach that would fall in line with his radical ideas. His xG model identified that Rayo Vallecano were performing particularly well in La Liga, so Brentford made an approach for their manager Paco Jémez. The Spanish coach turned down their advances and news of the approach filtered through to the media. When the story broke, Benham decided to go all-in on his revolutionary philosophy. He confirmed that Warburton's contract would not be renewed at the end of the season and braced himself for the inevitable backlash from the rest of the football world.

THE ROLE OF THE MANAGER

The statement Brentford eventually released following their defeat to Charlton, a week or so after the story of Warburton's impending departure initially broke in *The Times*, gives an insight into Benham's line of thinking. It is also uncannily prophetic in its ambitions for the club. Any team looking to overhaul its traditional structure in place of a forward-thinking, analytical philosophy would do well to use Brentford's 2015 statement as a template for how to go about enacting such change.

Firstly, the statement introduced the new management model that would be installed. A head coach would be supported by a Director of Football and a new recruitment model would be introduced, which would 'use a mixture of traditional scouting methods and other tools including mathematical modelling.'

By the 2030s we can expect the average managerial tenure to be less than a season.

The head coach would have an input, but not an absolute veto on transfer decisions. The press release also included a statement from Benham himself, 'It's difficult to implement change, especially when things appear to be going so well.' Notice the use of the word 'appear,' hinting that Benham's models didn't think the team were actually doing as well as the league table suggested. He continued, 'I am single-minded in my resolve that we can leave no stone unturned in our quest for sustainable football. Innovation, not increased funding, can be the only route to success for clubs such as ours, and I fully accept that innovation is never without risk.'

That final sentence perfectly encapsulated Benham's vision for the club. If David was going to defeat Goliath, he needed to choose a different weapon. He needed to innovate.

Both Matthew Benham's Brentford and John Henry's Liverpool realised the problems caused by the traditional role of the football manager, whereby ultimate decision-making power lies in the hands of one man. In the modern age of football, the lifespan of a manager is incredibly short. A head coach who doesn't show reasonable progress will often become maligned by the fanbase and end up sacked, while a head coach who outperforms expectations will often be poached by a bigger club. When English football resumed after World War Two, the average managerial tenure was around 2,500 days, the equivalent of 6.8 years. The lifespan of a manager in the modern day is about 20 per cent of that. A Premier League gaffer can now expect to serve for roughly 500 days, less than a year and a half. The length of managerial stewardship is ever-decreasing; if the current trend continues then by the early 2030s we can expect the average managerial tenure to be less than a season.

The traditional role of the manager is all-encompassing. They act as the figurehead for the club, handling everything from tactics, training, contracts, recruitment, agents and even the concerns of players' parents. Arsène Wenger was famously so intrinsically interwoven into Arsenal's fabric that banks demanded he sign a five-year deal with the club when loaning them the funds to build the Emirates Stadium. The manager is the central character in the soap opera that is football. In the media, in the stands, down the pub – wherever you turn, the sport belongs to the head coaches. Pundits and fans describe what the manager 'did' on the field, as if they themselves were playing the game. We say the manager won or lost. When things are going badly, who better to blame than the omnipotent leader of the club?

Despite their seemingly godly status, studies show that managers actually have surprisingly little impact on the success of their sides. The quality of a head coach tends to only influence the final standings of a club in the league table by about two points each season. The wealth and luck of a team play a much larger part in a club's success than the man in charge. Brian Clough seemed to understand this, once stating, 'Players win you games, not tactics. There's so much crap talked about tactics by people who barely know how to win at dominoes.' In other words, the most crucial role of the manager is filling the team with the best possible players. Signing quality players can be achieved via two avenues: either having more money than your opponents, or having a recruitment department who are betting at identifying hidden gems. The modern-day manager is having less and less involvement with player acquisition, as that responsibility pivots towards Directors of Football and Heads of Recruitment.

Moneyball is largely a tale of the struggle between Billy Beane's front office and the manager, Art Howe. Beane stated, 'There's this belief that a baseball team starts with the manager first. It doesn't.' Baseball's data revolution didn't originate from the bullpen. It originated from the back offices where Beane and his team of analysts went about tearing conventional wisdom to shreds. The manager's job isn't to guide long-term strategy or recruit players, nor should it be. The role of a head coach should be to squeeze every last ounce of ability out of each player. They should master the art of simplicity, breaking the sport down into fundamentals that players can easily understand. They should be expert communicators and relationship builders. Elite coaches will set the temperature and culture within a changing room, knowing when and how to push their players. They will proactively build and use the leaders within their squads and will always be thinking about what it takes to win. But even a manager who masters all these skills will only have so much influence on the final rankings of a team come the end of the season.

HOW TO MANAGE YOUR MANAGER

It's dangerous to place long-term strategic decision-making power in the hands of someone who is likely to leave your club within 16 months. Tottenham Hotspur offer a useful case study as to why. In late 2019, Spurs sacked Mauricio Pochettino, the manager who had led them to the Champions League final only a few months beforehand. His Spurs team were known for their aggressive attacking style and high-pressing game. The squad was built around this philosophy and was comprised of young, athletic players who were more comfortable in the opposition's half than their own. Tottenham's choice to replace Pochettino was the notoriously counter-attacking and reactive José Mourinho. The Portuguese preferred his teams to sit deep in their own half and contain the opposition before springing forward in transition. The squad that had been assembled and trained by Pochettino didn't fit the Mourinho style of play. The manager was left trying to force square pegs into round holes until he was eventually dismissed in April 2021, 17 months after his appointment. Tottenham were reportedly forced to pay an estimated £16 million as compensation for Mourinho's dismissal. And that's not to mention the cost incurred by the players the Portuguese signed during his tenure who may not have fitted with the next manager's playing philosophy. The hiring and firing of managers is a costly business.

Paul Barber, Brighton's CEO, recognises this. 'One trick we have to try and pull off is evolution,' Barber said. 'What we don't want to do is build a squad for this coach, and then he leaves and you have to build an entirely new squad for the next coach.' The football philosophy at smart clubs like Brighton is decided above the level of the coach, with the Director of Football, or their equivalent, meaning the club is fully aligned and isn't disrupted by managerial comings and goings. Brighton have a profile for the type of coach and playing style they want, and they recruit coaches who excel within that specific profile. This is an important distinction. They don't necessarily want the best coach, they want the best coach for their specific squad profile and style of play.

José Mourinho's appointment at Tottenham encapsulates many faults in the way managers are recruited by clubs. In the world of business, the search process for a new CEO usually takes four to five months. In football, a club usually finds a coach within a week of sacking his predecessor. Mourinho was hired the day after Mauricio Pochettino was given his marching orders. Football clubs rarely undertake appropriate due diligence on a potential managerial appointment. It would have been good practice for Tottenham to work behind the scenes for months, tracking and interviewing candidates and asking them to present their vision for the club, before eventually landing on Mourinho. Instead, the appointment was likely secured via a WhatsApp sent by the club to Mourinho's agent, offering him the job.

Once a manager is installed, a different challenge presents itself to the ownership. How do we assess their performance? Here, once again, Liverpool, Brighton and Brentford have trodden similar ground. If medium-term responsibilities such as recruitment and player contracts are taken out of the hands of your manager, how should you judge him? The key is to assess the performance of the team on the pitch. The manager is responsible for selecting the starting line-up, making the tactical decisions, and coaching the players, all of which impacts the on-field performance of the team. These are the things the manager controls, so this is what he should be judged on. At the beginning of the season, the powers that be must objectively measure the quality of the playing staff and set benchmarks they hope to achieve that campaign. Smart teams won't set targets like 'win the league' or 'avoid relegation'. In fact, they won't even look at the league table. The league table is a fibber; it doesn't accurately represent the ability of the teams. Remember how Benham's model ranked Brentford as the 11th best team in the division in 2014/15, even though the team managed by Warburton were sitting in the play-off places? John Henry and Tony Bloom will have a similar view based on their own models, which essentially ranks the teams according to their own xG performances. The managers will be assessed according to these alternative representations of the league table.

One of the most memorable moments of the *Moneyball* story is when Paul DePodesta, the Harvard University graduate and right-hand man to Billy Beane, sought to calculate how many runs the Oakland A's would need to reach the play-offs. Before the 2002 season, DePodesta reduced the upcoming campaign to a maths problem. He judged how many wins it would take for their team to reach the post-season play-offs (95). To win this many games, DePodesta estimated the A's needed to score 135 more runs than they allowed. (The concept that there was a stable relationship between season runs totals and season wins had been discovered by Bill James. A similar relationship exists in football: every goal is worth just over 0.5 points). DePodesta then used the past performances of the A's players to forecast that they'd score between 800 and 820 runs and concede between 650 and 670. From this he predicted they'd win between 93 and 97 games, enough to secure play-off entry. The A's ended up scoring exactly 800 and allowing 653 runs that season.

The owners of smart football clubs can recreate this equation using Expected Goals. Liverpool's analytics department will know exactly how much xG they need to produce and prevent in order to have a reasonable chance of winning the title. Figure 6.1 shows the average amount of xG that was created and conceded in each Premier League finishing position over an eight-year period. Creating around 80.00(xG) and conceding around 30.00(xG) over the course of a campaign should be enough to secure you the trophy. A team looking to avoid relegation should attempt to generate more than 40.00(xG) and concede less than 60.00(xG). Obviously this won't always be enough. Sometimes you'll win the league with less xG created or more xG conceded than you thought you'd need. Sometimes you'll suffer at the hands of extreme bad luck, as Brighton did during 2020/21. These situations, when luck rears its head in either direction, are when consultation of xG becomes even more important. A less shrewd club might have sacked Graham Potter when his team finished 16th in 2021. But Tony Bloom and his board were armed with xG data which allowed them to look past the results and recognise the true performance of Potter's team.

Figure 6.1 shows the average xG difference of each finishing position over the same period. Expected Goals difference works the same way as goal difference

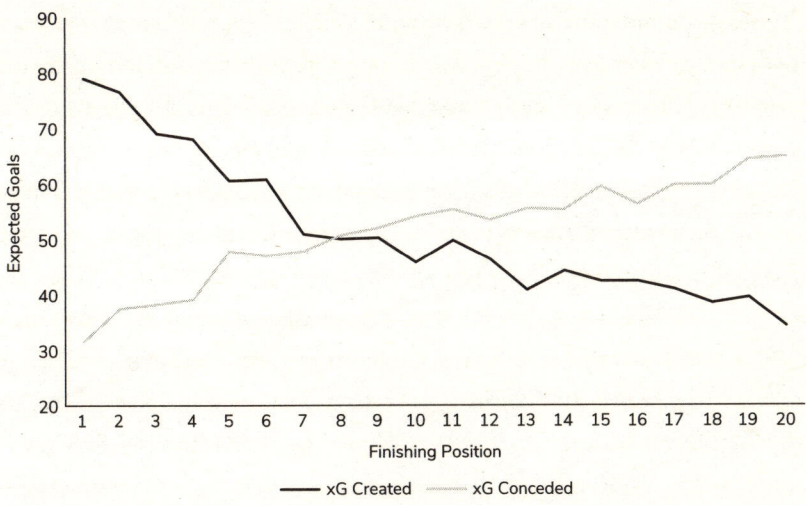

Figure 6.1: xG Created/Conceded vs Finishing Position, Premier League 2015–2023

– simply subtract the xG conceded from the xG created. The average xG difference for the Premier League winners from 2015–2023 was roughly +47.00(xG). That means a team expecting to win the title will need to create an average of 1.25(xG) more than their opponent in any given game. The average team who finished in 17th, narrowly avoiding relegation, accumulated around -19.00(xG). Any team losing by more than 0.50(xG) week-in, week-out will struggle to survive. The dominance of the top teams in the league over this period is immediately apparent in this graphic. Only the best seven sides tend to finish with positive xG difference, mirroring the top-heavy financial weighting of the league. Historically, the traditional 'big 6' teams have dominated the rest of the league, although this is starting to change with the emergence of challengers such as Newcastle United, Brighton, Aston Villa and others.

Analytically minded teams will assess the quality of their playing staff and work out where their team would be expected to finish along this chart come the end of the season, much like Paul DePodesta did at the Oakland A's. What should their xG difference be given the quality of the team compared to the other sides in the league? Measuring how the ability of a squad of footballers translates into xG, and therefore Expected Points, is more difficult than doing the equivalent in baseball. The fluidity of football makes it harder to model, but

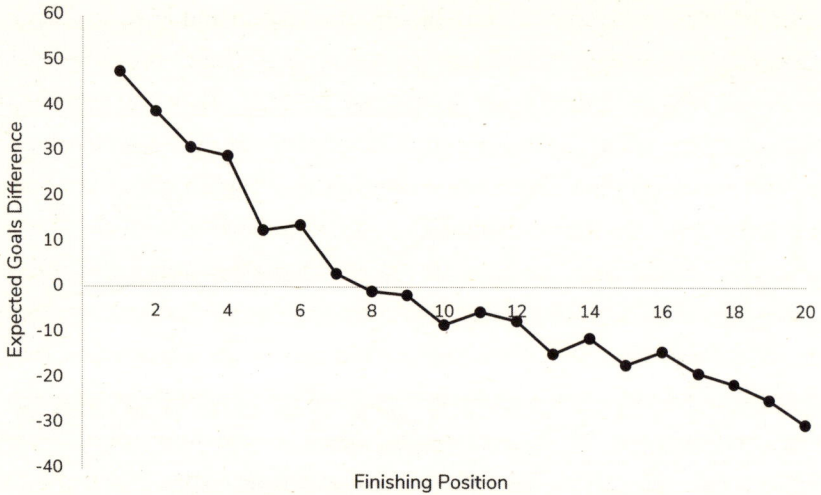

Figure 6.2: xG Difference vs Finishing Position, Premier League 2015–2023

companies like Starlizard and Smartodds are experts in this field. Brighton and Brentford will certainly have projections on where their teams will finish based on the players in their changing room. Using this, they can assess whether their manager is doing a good job or if he's underperforming expectations.

Thomas Frank won only two of his first 16 matches in charge of Brentford. The team slipped from sixth to 18th in the Championship at the start of his spell in 2018 and many fans called for the Dane's head. However, Brentford were actually hitting the xG targets that had been set by the board. They were creating lots of chances and conceding very few, but had been unlucky in many of Frank's opening matches in charge. Just as Matthew Benham's xG data had condemned Mark Warburton, it saved Thomas Frank. The owner stuck with him and Brentford went on a ten-match unbeaten run. Frank would eventually lead the Bees to their first ever promotion to the Premier League and to a top-half finish in their second season. Indeed, Lee Dykes, Brentford's Technical Director, confirmed that the club set such targets ahead of their first season in the Premier League. 'We knew how many goals we needed to score, and we knew how many goals we could concede at a maximum,' Dykes said. Although he wouldn't say so publicly, Brentford will actually be basing their targets on xG rather than actual goals. Performances rather than results.

John Henry of Liverpool, Tony Bloom of Brighton and Matthew Benham of Brentford all sought a different path for their clubs, one where short managerial tenures didn't upset long-term strategy. They all appointed a Director of Football, a fairly commonplace position in European football but one that was borderline revolutionary within the English game. The model they implemented in their organisations can be likened to a clock. The head coach is the second hand, responsible for the day-to-day running of the club: training the players, picking the team, developing tactics and so on. The Director of Football is the minute hand, responsible for medium-term strategies such as recruitment, player contracts and so on. The board, including Henry, Bloom and Benham at their respective clubs, is the hour hand, guiding long-term strategies and objectives (finances, budgets, expanding or building stadia, and so on).

Managers haven't historically cared about the long-term impact of their transfer decisions. There is no benefit to them thinking years ahead when they'll likely be in a different role by then. This is a crazy approach. The person who has access to the greatest expenditure in the business has no risk in the decision-making. Managers don't get a bonus or any other reward if the club turns a profit in any given season, which means there's no downside to spending recklessly in pursuit of short-term glory. Consistency, stability, and patience are ingredients that fall by the wayside.

Long-serving managers such as Sir Alex Ferguson or Arsène Wenger prove that perseverance serves dividends. It is well documented that Ferguson came under pressure at the beginning of his United career, but the owners stuck with him and he went on to create a dynasty at the club. Wenger was also intensely scrutinised at various points in his Arsenal tenure, but on each occasion the North London club's form dropped, it eventually reverted to normal and the owners were left thankful they'd stuck with the Frenchman. Billy Beane spoke admiringly of Wenger's frugal nature, stating, 'Arsène Wenger ran his football club like he was going to own it for one hundred years.' A more recent example of patience in practice can also be drawn from North London. Mikel Arteta came under extreme pressure from fans and the media after following consecutive eighth-placed Premier League finishes with a fifth-placed end to the 2021/22 campaign, meaning Arsenal didn't qualify for the Champions League in his first three seasons at the Emirates. The executives at the club stood behind him and in 2022/23 the team finished second – a position they'd only achieved once since 2004/05. When it comes to football managers, patience is a virtue.

ANTIDOTES TO MANAGERIAL DICTATORSHIP

A crucial component of Liverpool, Brighton, and Brentford's 'new world' was the implementation of a transfer committee. This caused problems with the managers of Liverpool and Brentford at the time. The move took the power of veto out of the hands of Brendan Rodgers and Mark Warburton and placed it in the dominion of a committee comprising several key stakeholders. At Liverpool, that included the manager, Head of Recruitment, Chief Scout, Chief Executive, President of FSG, Head of Analysis and the owner. The main responsibility of the committee is to identify transfer targets, consider their merits and value, decide which players to pursue and carry out negotiations with them. The committee would never sign a player without agreement from the manager, but similarly the committee could always say no to a player the manager brought forward.

The logic behind transfer committees can be traced back to a Fat Stock and Poultry Exhibition in the city of Plymouth in 1907. The exhibition was attended by Francis Galton, cousin of Charles Darwin and polymath originator of fingerprint identification and weather forecasts. Galton came across a large ox on display and contestants paying sixpence to guess the weight of the resulting meat from the poor beast after it had been slaughtered. He got hold of the 787 tickets that had been filled out and selected the middle (median) value of 547kg as the democratic choice. The estimate on every other ticket 'was condemned to be either too high or too low by the majority of voters.' The weight turned out to be 543kg, remarkably close to the median prediction. Galton wrote about his findings to the prestigious science journal *Nature* and titled his letter '*Vox Populi*' (voice of the people). This process of decision-making is now better known as 'wisdom of the crowds'.

Crowd wisdom has proven incredibly powerful in many remits, from forecasting elections to guessing how many jellybeans are in a jar. Any individual prediction-maker is likely to have a host of biases which affect their judgement. Collating numerous predictions will negate these biases and generally lead to a less noisy, more accurate view. If you ask one bettor to guess the probability of Arsenal defeating Manchester United in an upcoming game, his prediction will be riddled with a host of prejudices. If you ask the entire UK betting population, they will likely land on odds that accurately portray the chance of each event. The betting markets utilise wisdom of the crowds in price discovery and have proven possibly the best predictive tool available.

Sir Alex Ferguson's Manchester United may widely have been regarded as a one-man dictatorship, but even the great Scotsman harnessed wisdom of the crowds. Ferguson would regularly consult his players on transfer targets; who better to know about potential signings than the players who had faced them? Ferguson asked his centre-backs Gary Pallister and Steve Bruce for their opinion on Eric Cantona when he was debating a move for the Frenchman in 1992. The pair had played against Cantona when Leeds United had visited Old Trafford and both found him difficult to deal with because of the unusual positions he'd take up. United bought Cantona off the back of these recommendations. A year later the players unanimously told Ferguson that Nottingham Forest's Roy Keane was a top-tier talent, leading the club to fork out a British transfer record fee to secure him. Most notably, in 2003, Ferguson consulted United's defenders on the plane home from a friendly in Portugal. Sporting Lisbon's little-known teenage winger had been a constant threat, so Ferguson spent £12.2 million to secure the services of Cristiano Ronaldo.

Studies have found that even one person, on their own, can become a crowd. Think of this question: what is the distance between London and New York? As you thought about it, a number probably came to you. But it did not arrive in your head the same way as would your phone number or your age. You are aware that the number you produced is an estimate. It is not a random number – five miles or 10 million miles would clearly be wrong answers. But the number you produced is one in a range of possibilities that you would not rule out. If someone added or subtracted 2 per cent to your estimate, you probably wouldn't find the resulting guess much less plausible than yours.

Now, assume your guess is off the mark. Think about a few reasons why that might be. Which assumptions and considerations could you be wrong on? Was the first estimate too high or too low? Based on this new perspective, make a second, alternative guess.

Researchers have found that when subjects are asked to make estimates in this manner, the first guess is usually better than the second guess, but the best prediction can be taken by averaging both the guesses together. The process makes you think about information you might not have considered the first time around. (There are 3,461 miles between London and New York, in case you were wondering).

Averaging two guesses from the same person doesn't improve judgements by as much as asking for an independent second opinion. You can gain about $1/10^{th}$ as much value from asking yourself the question twice than you do from seeking an estimate from someone else. Interestingly, you can improve this figure to $1/3^{rd}$ if you let three weeks pass before asking yourself for the second estimate. Not bad for a technique that doesn't require any external support, and certainly a validation for the age-old advice to decision-makers, 'Sleep on it, think again in the morning.'

Group decision-making aggregates the judgments of multiple individuals to reduce unwanted 'noise.' But, if not careful, groups can add noise too. The decisions of groups can go in all sorts of directions based on factors that should be irrelevant. Studies have found juries to award different verdicts for similar crimes based on factors as trivial as whether the local sports team won that weekend. Another study found that general elections are swayed by the weather. If it's sunny, the public are more likely to be in a good mood and vote for the party currently in power. If it's rainy, they're more inclined to vote for the opposition party.

Who speaks first, who speaks last, who speaks with confidence, who is wearing red, who is sitting next to whom, who smiles or scowls or nods – all these factors, and many more, can affect group decisions. Every day, similar groups reach different conclusions, whether that involves determining environmental regula-tions, office closings, university admissions, communications strategies, or football transfers. Minor differences can lead one group to a comprehensive 'yes' and a near identical one to a firm 'no.'

Evidence of group noise in action can be found in an unlikely place: a large-scale study of music downloads carried out by Matthew Salganik and his co-authors. A control group was made consisting of thousands of visitors to a moderately popular website. Members of the control group could listen to and download songs by a number of new bands. The songs were vividly titled: 'Pink Aggression', 'Trapped in an Orange Peel', 'I am Error', 'Baseball Warlock v1', 'Gnaw' and so on. Users weren't told anything about what any other listeners had said or done. They were left to form independent judgments on which songs they liked and wanted to download. Salganik and his colleagues also made seven other groups, to which thousands of other website visitors were randomly assigned. These groups did have visibility over how many people in their particular group had previously downloaded each individual song. For example, if 'Pink Aggression' was gaining great traction among other members of their group, they could see that. So too if songs were receiving very few downloads.

You might expect the good songs to rise to the top every time, leaving the bad ones to sink to the bottom. The experiment was essentially re-running history eight times, so you'd predict each group to end up with identical (or at least similar) rankings. This is not what happened. The study found that songs which benefitted from early popularity did significantly better than those that didn't. In one group, 'Gnaw' would gain great acclaim and 'Baseball Warlock v1' would get hardly any downloads. In another group, the reverse would be true. The very worst songs, as defined by the control group, never ended up at the very top and the very best songs never ended up at the very bottom. But apart from that, anything went. The groups were incredibly noisy and that noise had a very particular driver: social influence.

Salganik and his co-authors followed this study up with a more mischievous one whereby they inverted the rankings of the control group. In other words, they lied about how popular the songs were. The website showed the least popular songs as the most popular, and vice versa. What did the users do? They started downloading more of the 'bad' songs and less of the 'good' songs. Perceived popularity and unpopularity bred more of the same. (The single exception was that the most popular song in the control group rose to prominence each and every time in the test groups – the inverted rankings could not keep the very best song down.) Similar effects can be found in the world of politics. The support of a political party or the popularity of a referendum proposal can be influenced by a small number of early adopters. An initial burst of popularity can be self-enforcing, while a proposal that attracts little support on its first day is essentially doomed.

Brentford have also sought to harness wisdom of the crowds by turning to fan forums to give them a helping hand in player reconnaissance.

These studies bear important implications on group judgements. Suppose a transfer committee is deciding whether or not to sign a player. If one or two advocates speak first, they might well shift the entire room in their preferred direction. The same is true if sceptics speak first. Similar groups might end up making completely different decisions based on the initiator of the conversation – the equivalent to the early music downloaders. Transfer committees can avoid this group noise by empowering each and every member to voice their opinion. If the committee are influenced by the social status of an individual member (say, the manager) then they're essentially passing the power back to him and undermining the formation of the committee. Ensuring a broad range of knowledge and experiences are present also combats social influence. The head coach will have his point of view, which will be independent from that of the head scout, which will be independent from that of the performance analyst, and so on and so forth.

Brentford have also sought to harness wisdom of the crowds by turning to fan forums to give them a helping hand in player reconnaissance. The evidence provided by thousands of eyeballs at multiple games is incredibly valuable. In a way, the club has turned its fanbase into one big scouting tool. Their supporters can provide quicker, more reliable feedback on a player than any one individual scout. But there is a potential hazard here. Consider a study Lev Muchnik and his colleagues carried out on a website that displays a diverse range of stories and allows people to post comments underneath, which in turn can be up voted or down voted. The researchers automatically and artificially gave certain comments an immediate up vote (the first vote the comment received). You might not expect a single up vote to matter, given the hundreds of thousands of visitors and ratings. You'd be wrong. After seeing an initial up vote (which, remember, is entirely artificial) the next viewer became 32 per cent more likely to give an up vote to that comment. When fan forums are abuzz with praise for a potential new signing, you must be aware that the player may have gained favour, in part, by the equivalent of an early up vote. Wisdom of the crowds works best if all the opinions are independent and not influenced by one another. Fan forums and social media feeds can act as echo chambers. A small portion of vocal fans can cause a cascade of similar opinions on the quality of a player, in a similar way to the music downloads example. The way groups move in the direction of people, products, and ideas can be greatly influenced by noise.

Clearly it's not enough to simply assemble a transfer committee and let the wisdom of the crowds do its thing. Crowds, too, can generate undesirable noise. That's where data comes in. Statistical analysis provides an objective view that is largely stripped of unwanted noise. Data doesn't care who spoke first, who spoke loudest or who was wearing red. It doesn't feel the pressure of social influence. Rasmus Ankersen, who was installed as a Director of Football at Brentford after Mark Warburton's departure, said, 'We re-designed the club based on one question: what would a football club look like if it had no human eye and ear?' Brentford's mantra is to 'distrust your eyes.' They lie to you. What you see is blinded by bias and emotion. Strip these detrimental factors from your analysis and you'll be left with much clearer vision.

Sports teams are even taking power out of the hands of the manager during the game. The Brentford manager and his coaching staff are wired in to 'tactical statisticians' who sit in the stands during the game. These analysts track the running data, allowing them to spot if a player is becoming tired and should be substituted. They have a better view of the pitch, meaning they can advise on any formation changes from the opposition and provide tactical guidance. They also track the xG being accumulated by either side, which offers an impression of the flow of the game and the dominance of either team. The analysts offer various strategic suggestions based on how the alterations will affect the team's odds of success. The data revolution in professional sports is a revolution in probabilistic thinking. Teams are trying to work out which in-game events, big or small, contribute to the chances of winning. They're trying to follow the most optimal path along the decision-making tree as they can. Billy Beane famously never watched the Oakland A's play. He'd spend the matches in the stadium gym or watching local baseball teams play instead. 'All they provide me with is subjective emotion, and that can be counterproductive,' Beane said. If a general manager (the person in charge of trading players) feels this way, imagine being a coach who has to make objective decisions in real-time. In-play advice from analysts is just another way in which power is taken out of the hands of the manager and put in the hands of objective stats.

Starlizard and Smartodds, founded by their owners long before they even took control at their respective clubs, act as the central data hubs for Brighton and Brentford. While other teams pay for the privilege of accessing stats that any other club can pay to access from companies such as Opta, Brighton and Brentford have their own pools of unique, tailored metrics, far more powerful than any available for purchase. This gives them a head start on other teams and an advantage that is difficult to replicate. These companies are the brains behind the operation. They are the computing systems that allow the clubs to make decisions. Brighton and Brentford are using iMacs while other teams are still stuck on iPod nanos.

But the Fenway Sports Group case study gives other clubs cause for optimism. Liverpool brought in analysts, devised their own models, and, crucially, set the club up within a framework that allowed analytics to come to the fore. The way they shaped their organisation also allowed consistency to take pride of place. Sport is a revolving door, and it speaks to the quality of the organisational structure that most of the front office and management team stayed intact throughout Jürgen Klopp's reign. The ownership struck a nuanced balance of not being too involved in the day-to-day operational minutiae and efficiently delegating to their team of best-in-class operatives, while still driving all strategy and performance initiatives. That's a difficult line to tread, especially in sport, but the FSG principals navigate it better than most.

Some sections of the Liverpool fanbase might not be so quick to praise Fenway Sports Group in their running of the club. Questions have been levelled at John Henry and the ownership group regarding a perceived lack of spending in the transfer market. Indeed, FSG seem to treat Liverpool as an investment and certainly spend more carefully than rivals such as Chelsea and Manchester United. When Liverpool are outbid for players such as Moisés Caicedo, who chose the Blues over the Reds in the summer of 2023, some sections of the Scouse fanbase become emotional and perhaps a little overreactive. But whatever your views on how deep FSG should be reaching into their pockets, there's no doubt that Liverpool's record under their stewardship, both in terms of on-field success and off-field revenue driving activities such as transfers and commercial growth, has been outstanding.

FSG showed that you don't necessarily need advanced modelling capabilities or ground-breaking metrics to structure your club in a way that redistributes power and allows for a more logical, coherent approach to the sport. Once a controversial position, the value of a Sporting Director or equivalent is now accepted by most clubs. All 20 Premier League teams currently have a Sporting or Technical Director in place. While the role itself can vary, installing this position in your club is a critical lever for ensuring sustained success on and off the pitch.

The three traditional roles of the manager are coaching, recruitment and tactics. The advent of the Director of Football role means recruitment is being carved out from the list of managerial responsibilities. Might we also see the other two major duties devolved into specialised roles in the near future? Perhaps clubs will start employing a 'Head of Coaching' who is responsible for technical training with an emphasis on player development. The advancement of a player's physical, mental, and technical ability is a long-term objective which could be considered independent from the team's results on the pitch. Similarly, clubs could hire a 'Head of Tactics' who would be in charge of team selection, formation, playing style and set-piece routines. Most managers would slot more easily into this role than the other two. What sets Pep Guardiola and Jürgen Klopp apart isn't their ability to recruit undervalued players or coach them into passing better, but their capacity to improve teams by moving the various chess pieces around the board.

Separating these roles out would leave clubs less vulnerable to managerial changes. We have seen how Brighton and Brentford have been able to consistently sign great players because their recruitment department has endured, even when their head coaches have been replaced. If a club who separate the roles were to enter a crisis period and needed to shake things up, they could theoretically replace the Head of Tactics with no impact on transfer strategy or player development. They could isolate the specific area where things had gone wrong (for example, the tactical set up of the team) and make a change without affecting the other areas of the club which had been performing well (recruitment and coaching). The same principle would apply if any of the other areas were underperforming.

Another benefit of such a structure would be the ability to employ the highest-quality coaches in each of the devolved units. Usually when hiring a manager, clubs have to choose between a coach who is either great tactically, great at developing players or great at recruitment. It's rare to get a manager who is strong in all three remits. By splitting out the responsibilities, you could theoretically employ the best Head of Tactics sitting in the dugout on the weekend, the best Head of Coaching on the training field, and the best Head of Recruitment crunching the numbers in his office. You could have the best of all worlds, a holy trinity of management staff who are all highly specialised and effective in their specific roles.

Analytically minded clubs have sought to restructure their organisations in an attempt to reduce bias, noise and human emotion from the equation. These teams have diversified the spread of power. The heavy burden previously weighing entirely on the manager's shoulders is now being shared among experts, executives, Directors of Football, transfer committees, technical performance analysts and advanced mathematical models. Human intuition is gaining greater support from data. Traditionally, the role of the manager was to be the single pillar supporting the entire infrastructure of a building. Clubs have recently begun to build other pillars to help share the responsibility and carry the weight. Before, the whole roof would cave in if the managerial pillar collapsed. Now if this pillar disintegrates, the Director of Football pillar, Head of Recruitment pillar, and a host of other pillars maintain the solidity of the structure and a new managerial pillar can slot seamlessly in. The structure remains intact and an expensive, time-consuming rebuilding process is avoided. Any club not ripping up their current model and emulating Liverpool, Brighton and Brentford is being left behind. It's easy to forget the resistance these clubs were met with when the structural changes were first brought in. Fenway Sports Group and Matthew Benham were ostracised by the media, their radical organisational shakeups branded as insanity.

Indeed, genius usually looks a lot like insanity at first.

CHAPTER SUMMARY

- The traditional role of the manager is all-encompassing. They have historically been responsible for tactics, recruitment, training, and a host of other tasks. Smart clubs are splitting the role between a cohort of specialists and using wisdom of the crowds to eradicate individual biases from decision-making.
- Transfer committees and Directors of Football are becoming increasingly prevalent as clubs appreciate the need to take power out of the hands of the manager.
- Forward-thinking teams are becoming better at recruiting managers, as well as measuring their performance once they're installed in their role. These clubs will ultimately win more football matches.

7

THE RECRUITMENT DEPARTMENT

HOW TO UNEARTH HIDDEN GEMS

'Player recruitment is the most important application of analytics by a factor of 10'

Dr Ian Graham, former Liverpool FC Director of Research

Deepak Ravindran and his wife Emilie Vanpoperinghe are in the wonky vegetable business.

It all started when they came across a misshapen tomato while on holiday in Portugal. The vegetable looked ugly but tasted delicious, which got the couple wondering why fruit and vegetables in the UK always looked appealing but never tasted as good. A short amount of research revealed that the food production industry creates huge amounts of perfectly edible produce, only for supermarkets to turn 30-40 per cent of it away because it doesn't meet 'consumer standards'. One-third of food produced for human consumption is lost or wasted globally, the equivalent of 1.3 billion tonnes per year. The UK alone sends 90,000 tonnes of produce to landfill, but this food is never recorded as supermarket waste because it's rejected before it hits the shelves. Ravindran reckoned a lot of consumers would be happy to buy fruit and vegetables that didn't meet supermarket standards. He staged a trial, buying misshapen produce from local suppliers and delivering them to houses around his home in London. The trial was successful enough that Ravindran and Vanpoperinghe pooled their savings and launched their own company: OddBox.

Football clubs can learn a thing or two from the repurposing of forked carrots, conjoined apples, and scarred aubergines. Just like supermarkets turn away tasty produce, football clubs too easily dismiss perfectly good footballers. The market for players is massively inefficient and can be exploited by those who don't mind buying the odd misshapen tomato. There's a great deal wrong with how people judge the ability of footballers: some traits are drastically overvalued, while others are hugely undervalued. Smart clubs are using analytics to identify oddly shaped fruit which they can buy for a pittance but still tastes just as flavoursome, if not more so, than beautifully formed alternatives.

The size of transfer fees being paid for footballers in the modern day has occasionally prompted a sense of unease, particularly from those outside the sport. Many believe that the relentless growth of transfer sums is unsustainable. However, analysis shows Premier League clubs, who account for roughly a quarter of global transfer expenditure, have consistently spent around 15 per cent of revenue on transfers over the last few decades. As budget lines have swelled, so have transfer fees. These figures indicate the market isn't as 'crazy' as some people believe – the expenditure on playing staff is merely mirroring the finances flooding into the game from TV revenue, increased commercial deals and billionaire owners. The numbers also reveal how much the purchase or sale of a player can impact a team's financial state. Clearly getting these decisions right is crucial to the success of a club, particularly those operating on restricted budgets.

The four key revenue drivers for teams are broadcasting rights, match-day ticketing, commercial activities, and the transfer market. The only way to materially grow the first two in this list is to achieve promotion and the third is closely aligned with the size of the fanbase, which is also largely dependent on the performance of the team on the pitch both historically and in the present day. The fourth of these, player trading, is the only one fully in the dominion of the club. Every team has the ability to become smarter and make better recruitment decisions. In a sport where 75 to 85 per cent of the explanation of results can be put down to the money a team spends on their first-team squad, clubs who can learn to control the transfer market gain a massive advantage. Analytics has a massive part to play in unveiling the true ability of footballers and helping teams bring in vital revenue through acquiring, developing and selling players.

But player trading is difficult. Dr Ian Graham, former Director of Research at Liverpool and an integral part of their recent success, estimates that half of all transfers fail. Graham outlined the six ways a new recruit might flop:

1. The player isn't as good as you thought.
2. The player doesn't fit your style.
3. The player is played out of position.
4. The manager doesn't like the player.
5. The player has injuries or personal problems.
6. A player already in the squad ends up being better.

There are obviously ways to mitigate the risk of these factors, but even if you're 90 per cent sure of success in each of these areas, that means a transfer only has a 53 per cent chance of succeeding ($0.9 \times 0.9 \times 0.9 \times 0.9 \times 0.9 \times 0.9$). The difficulty in getting transfers right, combined with the importance of transfers on the success and financial standing of a team, means making smart recruitment decisions is essential. 'Player recruitment and retention is the most important application of analytics by a factor of 10,' said Graham. 'It's where the action is at.'

Data can help accurately benchmark the ability of the teams you wish to buy players from, making you more comfortable that their talent can make the step up. It can prevent us from overpaying for unimportant attributes or qualities. It can help us better understand age profiles and identify depreciating assets. Even simple data is often overlooked when deciding whether to buy or sell a player. For example, 'minutes played' is a metric hardly ever mentioned when discussing the quality of a footballer. How often is a player injured? Availability is both one of the most important and one of the most forgotten stats. We might salivate over a player generating $0.90(xG/90)$, but this output is next to useless if he is only able to play 90 minutes each season. And buying talent is only half the story. Spotting when to sell a player and extracting value for your deflating assets is equally as important as finding bargains.

Brighton & Hove Albion are perhaps the most well-known masters of the transfer market. Their market-leading recruitment methods based on the chance-creation data collected by Starlizard have allowed them to consistently identify hidden gems. A host of players, many of whom are outlined in Figure 7.1, were undervalued prospects who were picked up by Tony Bloom's company's model and signed on the cheap. The club's bold approach to selling players has then allowed them to make large profits, before reinvesting that money in a new array of undervalued talent. The eight players in the table

Player	Purchased			Sold			Profit
	Year	Club	Fee	Year	Club	Fee	
Moisés Caicedo	2021	Independiente del Valle	£24m	2023	Chelsea	£115m	£91m
Ben White	2018	Brighton U23	£0	2021	Arsenal	£50m	£50m
Marc Cucurella	2021	Getafe	£15.5m	2022	Chelsea	£63m	£47.5m
Alexis Mac Allister	2019	Argentinos Jrs	£7m	2023	Liverpool	£35m	£28m
Yves Bissouma	2018	LOSC Lille	£14.5m	2022	Tottenham	£35m	£20.5m
Robert Sánchez	2013	Levante	£0	2023	Chelsea	£20m	£20m
Leandro Trossard	2019	Genk	£13.5m	2023	Arsenal	£27m	£13.5m
Dan Burn	2018	Wigan	£3m	2022	Newcastle	£13m	£10m
			£77.5m			£358m	£280.5m

Figure 7.1: Brighton & Hove Albion's Transfer Market Activity

were bought for a combined £77.5 million and sold for nearly five times that figure. The £280.5 million profit made on these players represents a windfall roughly 70 per cent greater than a team tends to make from winning the Champions League. This graphic only goes some way to outlining the Seagulls' transfer-market brilliance. A host of talented players are still currently playing for the team at the time of writing, including the likes of Evan Ferguson, Pervis Estupiñán, Kaoru Mitoma, Julio Enciso, Simon Adingra and Facundo Buonanotte. These players are all now valued several times higher than the fees the club paid for them.

Bloom's xG data showed that the 5-0 match had actually been very close.

In the summer of 2018, Tony Bloom was looking to acquire another team. His requirements were that it should be in Europe but easily accessible from London, relatively inexpensive, have the potential for growth, and possess a welcoming fanbase. They landed on a small Belgian club called Royale Union Saint-Gilloise. Its stadium sits on the corner of a picturesque park and houses just 9,000 fans. The club resided in the second tier and was dwarfed by close neighbours Anderlecht, the Belgian giants who boasted 34 league titles, nine cups and five European trophies. Royale Union Saint-Gilloise had spent the previous half-century bouncing around the lower leagues and had a smaller budget than many amateur teams. Bloom spotted a club which was surviving, and hoped his methods could make it thrive.

Recruitment was always going to be the key to their success. The club suddenly had access to Starlizard's pool of data and analytical services, the same way Brighton did. Success didn't come immediately to Union. They made mistakes along the way, but the adoption and refining of the recruitment methods outlined in this chapter saw the club's fortunes change on its head. The club were promoted back to the first division in 2021. At the beginning of their first season back in the top flight, Union were handed odds of 500/1 to win the league. In the previous season Anderlecht had beaten them 5-0 in the cup. 'The effect that it had was it put a load of teams off our players and off our team,' said Alex Muzio, Bloom's business partner and chairman of Royale Union Saint-Gilloise. 'They thought we were bad.'

This turned out to be a blessing in disguise as it meant Union could keep their squad intact. In reality, Bloom's xG data showed that the 5-0 match had actually been very close. Anderlecht had scored with virtually every shot they'd taken. Starlizard's algorithms reckoned Union were much better than the pundits, bookmakers and fans were giving them credit for. They were right. Royale Union Saint-Gilloise ended up five points clear of second-place Club Brugge and 13 points clear of third-place Anderlecht in the regular season. The transfer policy espoused by Bloom has succeeded at various clubs in different levels and different leagues. Their blueprint is one every club should be copying.

Even clubs at the other end of the financial spectrum to Brighton and Royale Union Saint-Gilloise require constant transformation of their playing staff. For a while, retaining the Premier League title proved impossible – no club managed to do so in the nine seasons between 2009 and 2018. After that spell, Manchester City won it five times in six seasons. Yes, the depth of their owners' pockets played a large part, but so did their attitude to refurbishing the squad after every season – even (and perhaps especially) after successful ones. Pep Guardiola recognised that if you stand still for too long, you will be caught up. He's allowed talented players like Gabriel Jesus, Raheem Sterling, Ferran Torres and Leroy Sané to leave the club, bringing in hungry new talent to replace them. He's also been willing to let go of older players like Sergio Agüero, Vincent Kompany and İlkay Gündoğan sooner than other clubs might have done, not allowing himself to be nostalgic and keep players simply because of their past achievements. What matters is their quality in the present moment. The previous side to retain the Premier League title before Manchester City were Manchester United under Sir Alex Ferguson in 2008/09, a team who displayed a similarly aggressive approach to player trading and a similar desire to constantly revamp the playing staff. Ferguson was always on the look-out for signs of decay and took no prisoners. Players including Mark Hughes, David Beckham, Paul Ince, Jaap Stam, Ruud van Nistelrooy and Roy Keane were all released with surprising haste. Premier League winners cannot rest on their laurels; the teams chasing them certainly won't.

The transfer market, just like any other market, is about buying and selling. Football fan forums, social media feeds and gossip columns tend to focus on the former of these activities. Supporters get far more excited by potential arrivals than potential outgoings, as shown by the modern trend of 'announcement videos' that have spawned in recent seasons. 'In the know' journalists like Fabrizio Romano have found huge success by being the first to reveal who clubs are looking to sign. There is something of a journalistic arms race to be the first one to announce new arrivals to hordes of baying supporters. All the hype centres around the procurement of footballers. But selling players at the right time and the right price is equally as important as purchasing new ones. Let's first turn our attention to how intelligent clubs dispense of their talent.

GOALSCORER INFLATION

Michu had reached god-like status. Arriving at Swansea City for just £2 million from cash-strapped La Liga side Rayo Vallecano in 2012, the striker was previously unheard of in the English game. Michu announced himself on his Premier League debut against QPR by scoring two goals and assisting another. He ended up scoring 22 goals in the 2012/13 season, including one in Swansea's victorious League Cup final against Bradford City. His performances even led to a first cap for the Spanish national side. Following his break-through campaign, Michu was linked with a move to some of the giants of the sport. Arsenal reportedly submitted a £25 million bid, while Liverpool also enquired about his services. But Swansea refused to accept any offer less than £30 million, a price no club was willing to match.

Michu's second season in South Wales started less well. A handful of mediocre performances were followed by an ankle injury. When he returned to the field he couldn't replicate the form he had shown in 2012/13. He was loaned out to Napoli, where he flopped. The Siren's Song, one of Napoli's most popular blogs, described Michu as 'just plain bad'. The Spaniard is now the unfortunate poster boy for a group of players known as 'one-season wonders'.

Michu fell victim to what is known as 'regression to the mean,' another statistical phenomenon which originated with Francis Galton. When he wasn't eliciting the wisdom of the crowds to measure the weight of oxen, Galton was investigating the way personal characteristics change between generations. Using their parents' height, can we predict an adult offspring's height? Galton had the classic Victorian scientist's obsessive interest in collecting data. He compared 465 sons' height to their fathers' height and made an interesting discovery. Tall fathers tended to have smaller but still taller-than-average sons, while short fathers tended to have taller but still small-than-average sons. Galton called this 'regression to mediocrity,' but it's now known as 'regression to the mean'.

If Michu's 2012/13 campaign was a 6-feet-6-inch father, his 2013/14 campaign was always likely to be a shorter son. The Spaniard had enjoyed an extraordinary season and was always likely to regress to the mean. In fact, even his performances at Swansea during his miracle campaign hinted at an eventual decline; Michu wasn't a dazzling performer, his work-rate off the ball was poor and he amassed few assists. But the footballing world became engrossed by him because of the sheer number of goals he was scoring. Goals are the ultimate currency in football, but their randomness can lead to players becoming grossly and irrationally overhyped. Arsenal and Liverpool dodged a bullet by not agreeing to sign Michu for the £30 million Swansea were demanding.

Players who score lots of goals one season aren't necessarily likely to repeat the feat the next. Figure 7.2 shows the average regression for each 'bucket' of goals scored in a Premier League season by a player between 2004 and 2023. For example, players who scored 15 goals in a campaign went on to score an average of 7.1 goals the next season, a regression of -7.9 goals. The sample includes all 112 players who scored 15 goals in a season and went on to play at least 900 minutes the next campaign over this 19-year time period. Of these 112 players, only 17 went on to improve on their scoring tally the following season (15 per cent). The average regression for a player who scored 15 goals or more was -6.2 and overall the cohort of 112 players scored 669 fewer goals the season after their 15+ campaign.[8]

---

[8] Note: no player scored 28 goals during a Premier League campaign over this period, so the number has been omitted from the chart.

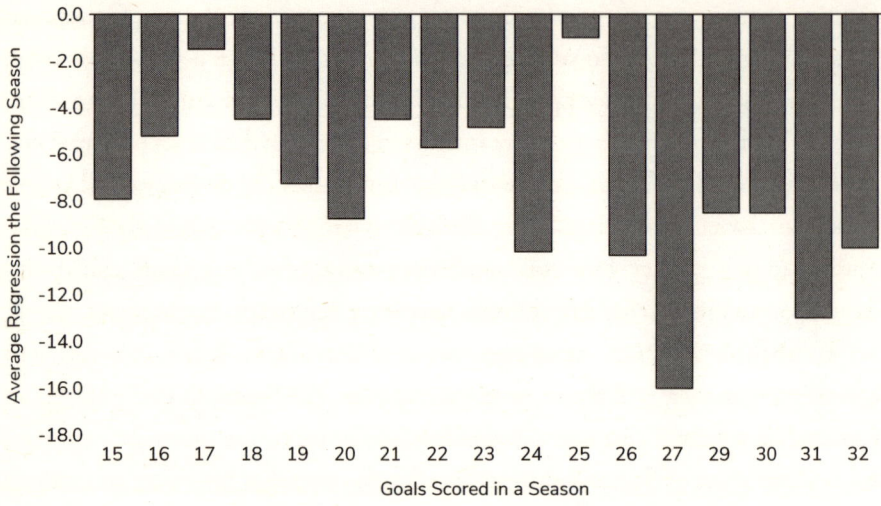

Figure 7.2: Regression in Goalscoring, Premier League 2004–2023

The majority of the 17 instances when an improvement was made season-on-season were when players approached the peaks of their powers. The best progression was made by a 22-year-old Cristiano Ronaldo, who went from 17 goals in 2006/07 (he was 22 at the end of this season) to 31 in 2007/08, while a 26-year-old Luis Suárez went from 23 in 2012/13 to 31 in 2013/14, and a 25-year-old Sergio Agüero went from 17 in 2013/14 to 26 in 2014/15. Harry Kane improved his goals tally for three consecutive seasons from 2014/15 to 2017/18 as he became one of the world's most prolific marksmen. Kane also improved season-on-season from 2018/19 to 2019/20 and then again to 2020/21. He's 'progressed' rather than 'regressed' six times over the sample period, meaning Kane alone accounts for more than one third of the player seasons in this sample to have improved year-on-year. If you strip Harry Kane from the study, only 11 of the remaining 104 campaigns (11 per cent) actually show improvement versus the prior season.

Michu's lapse from 18 goals in 2012/13 to two in 2013/14 represents a regression of -16, the joint-third worst figure in the study. The biggest drop-off was recorded by Didier Drogba, who accumulated an outstanding 29 goals in 2009/10 before notching 11 the following season. Regression does seem to hit harder the further to the right of the graphic you venture. The more goals you score one season, the bigger decline you're likely to experience the next. Players who score record-breaking numbers of goals are likely to drop off more severely than mid-tier scorers. The chart gives credence to the old adage, 'the bigger they are, the harder they fall'.

Scoring regression can be caused by a number of factors. Availability might play its part, as was the case with Michu. A player who scores 15 or more goals in a campaign was likely able to avoid injuries and play a lot of minutes. Such fortunate fitness is not necessarily repeatable – it only takes one mistimed tackle to cut a striker's season short. An individual player can expect to be injured once every 28 matches, while the average recovery time is around three weeks. Michu played over 3,000 minutes in his breakthrough season at Swansea, but only managed around 1,300 the next. Fitness is very much in the hands of chance, and Lady Luck can be a fickle mistress.

Players who score record-breaking numbers of goals are likely to drop off more severely than mid-tier scorers.

Another reason a player's goalscoring output might regress is if their team is worse than the prior season, a fact Steven Gerrard can testify to. The Liverpool midfielder bagged 16 goals as his team finished second-place in the Premier League in 2008/09. Liverpool struggled for form the next season, creating fewer scoring opportunities and finishing 23 points worse off. The chances dried up for Gerrard, who notched seven fewer goals than the previous campaign. Forward players depend on teammates to provide them with scoring opportunities and the quality of service they receive depends on the dominance of their team. Most of the forwards in the sample played for clubs towards the top of the league table. But just as players can enjoy anomalous seasons, so can teams. When an over-performing team regress to the mean, it's likely that the scoring output of their players will do too.

Finishing consistency is also a big factor in the mean regression of strikers. We'll see later on how the repeatability of 'clinicalness' is questionable. A striker who drastically outperforms xG one season is likely to crash down to earth the next. Jamie Vardy scored 23 goals from 19.25(xG) on his way to the Premier League Golden Boot in 2019/20. Perhaps the goalkeepers he faced let more in than they should, or the ball bobbled favourably when he took his shots. Whatever the case may have been, luck flipped on its head the following season. Vardy netted 15 goals, eight fewer than the prior campaign from an almost identical xG (19.51).

Sometimes regression can be aggravated by raised expectations from the prior season. Pierre-Emerick Aubameyang scored 22 goals from 16.10(xG) for Arsenal in 2019/20 – about six goals more than expected given the chances presented to him. He earned multiple plaudits as a result but wasn't so lucky the next season, scoring 10 goals from 10.31(xG), an overall drop-off of 12 goals from the prior season. This comparatively dry spell in front of goal led to criticism from some corners of the Arsenal support, having had their own expectations raised by his free-scoring campaign just gone. Perhaps this added weight on his shoulders contributed to the breakdown in relations between him and Mikel Arteta. Aubameyang's aggressive goalscoring regression might have been the catalyst for his eventual exit from the club a year or so later.

The transfer of a player to a new team has also been known to cause a significant decline in scoring output. Fernando Torres netted 81 times in 142 appearances for Liverpool before moving to Chelsea for a reported £50 million, the most expensive British transfer fee ever paid and the sixth most expensive transfer overall at the time. Expectations were high when Torres joined in January 2011, but he only scored one goal in the 14 Premier League appearances he made before the end of his first season. Chelsea's style of play was completely different to what he had experienced at Liverpool. He had succeeded in a counter-attacking system where Steven Gerrard or Xabi Alonso would release him with defence-splitting passes. Chelsea, on the other hand, were more of a possession team and Torres would rarely get a chance to run off the defender's shoulder. The Spaniard's skill didn't lie in a good first touch or strong link-up play, he was all about raw pace and lethal finishing. His main attributes were not brought into play at Chelsea which meant his scoring output suffered. Torres went on to accumulate 45 goals in 172 appearances for the Blues before leaving the club.

So what does this all mean for how teams approach the transfer market? Well, clubs who spot signs of regression can cash in before their assets begin to depreciate. As a basic rule, a striker who has an incredible scoring season is unlikely to maintain that level of performance. But most decision-makers don't realise this. They get swept up in a player's success story, the buzz, and the hype, which leads to the player's value becoming inflated. Smart clubs will take advantage of mean regression, looking to sell players while their stock is high. These clubs' approach to buying and selling players is similar to how a trader might approach the stock market. Each footballer represents a value, a sum of money that they are worth. It's obvious to buy undervalued players, but clubs often forget about the other side of the coin: when one of your own players has become overvalued by the market, sell them.

How many points do you think Harry Kane was worth to Tottenham per season during his peak years? That is, how many points would Spurs lose if they sold Kane? A room of sporting directors was once asked this question. The answers tended to range from 10 to 15 points, with some reckoning as high as 25. Some back-of-the-envelope maths can help us answer the question. In most major leagues, a goal scored (or conceded) is worth just over 0.5 points over the course of a campaign. Kane averaged 25 goals per season between 2014 and 2022, so it stands to reason that he contributed around 13 points per campaign. But, of course, Kane's absence would have been filled by another player. We might expect an average striker who comes in to fill Kane's boots to notch about 13 goals per season – a contribution of around seven points. Therefore the 'cost' to Tottenham for Kane's absence would be around six points. This is certainly a significant amount, but nowhere near the numbers being thrown around among the room of sporting directors.

Our answer can be corroborated using a different method. If a modern-day title-winning team earns around 90 points in a season and a relegated team collects 35 points, we know the players in a leading team are worth around 55 points more than the players in the weaker team. That works out to an average of five points per player in the starting line-up. Whichever way you look at it, individual footballers are rarely worth more than six points per season to their team.

Studying the relationship between performance and income in the Premier League reveals that a decrease of six points and the resulting drop down the table would cause an average long-term revenue reduction of roughly 18 per cent. Tottenham's average revenue in this period was an estimated £460 million, meaning Kane's absence would cause them a long-term hit of £83 million. Clubs will no doubt have more sophisticated means of assessing player value based on a range of different variables, but this quick calculation gives some indication as to where Kane benchmarks. Clearly this implied financial impact is substantially less that the transfer fees paid for other world-class players such as Philippe Coutinho (£142m by Barcelona), Jack Grealish (£100m by Manchester City) and Romelu Lukaku (£97.5m by Chelsea), all of whom it could reasonably be argued have never reached the heights or consistent performance output of Kane. The Tottenham striker's implied value is in a similar ballpark to what Manchester United paid for players like Paul Pogba (£89m), Antony (£82m) and Harry Maguire (£80m). Few would argue these players are on a par with Kane in terms of ability or value.

The key takeaway is that clubs tend to overvalue the impact of players and overspend on new signings. Consider the sporting directors who thought selling Kane would cost Tottenham 25 points per season. Following the same line of analysis, it's implied that these executives thought Kane was worth £345m. Clearly they either misunderstood the exchange rate between goals and revenue, or didn't account for the fact that another striker would come in to take Kane's place. The upshot of the overvaluing of players is that teams tend to receive much greater transfer compensation than they deserve. The actual fees paid for players consistently exceed the true value of their worth. This makes selling your players an incredibly worthwhile endeavour, especially as there's a strong case to be made that few signings over €100 million have actually been a success. The below list includes every €100 million transfer made before 2023:

- **Eden Hazard** – Real Madrid (€100m): The move from Chelsea started off poorly when he arrived for his first training camp overweight. He failed to start more than 20 league games in a season at Madrid and didn't score a league goal between May 2021 and his retirement in October 2023 at the age of just 32.

- **Paul Pogba** – Manchester United (€105m): The most memorable thing about Pogba's United spell was his rivalry with Graeme Souness. He did have his moments, but they tended to be more bad than good.
- **Ousmane Dembélé** – Barcelona (€105m): A great player who has been injured around 35 per cent of his time at Barcelona. If the club had known what they were going to get, it's highly unlikely they'd have made the signing.
- **Romelu Lukaku** – Chelsea (€115m): A move which seemed perfect. A highly clinical finisher moving to a team which was creating and squandering plenty of chances. But the transfer was a catastrophe. Lukaku scored eight Premier League goals before being shipped back to Inter Milan.
- **Jack Grealish** – Manchester City (€117m): A tricky one to judge given City's success, but for the price you'd probably want more than eight goals and 10 assists in his first two seasons.
- **Antoine Griezmann** – Barcelona (€120m): Never quite found the form he showed at Atlético Madrid and played second string to Lionel Messi and Luis Suárez. Admitted his struggles before being shipped back to Madrid.
- **João Félix** – Atlético Madrid (€126m): A good player who didn't fit the Diego Simeone system. Won one trophy in four years and didn't manage more than 15 goals plus assists in any of them.
- **Phillippe Coutinho** – Barcelona (€145m): Another undeniable catastrophe. The club were looking to sell him a year after he joined, having been nowhere near the standard he showed at Liverpool. Left the club to join Aston Villa for around €19 million.
- **Kylian Mbappé** – Paris Saint-Germain (€180m): Although arguably the best player in the world, has Mbappé drastically altered the club's fortunes? They likely would have continued to win Ligue 1 without him, while his presence hasn't allowed them to win a Champions League.
- **Neymar** – Paris Saint-Germain (€222m): The same point as Mbappé could be made, with the additional points that Neymar cost a lot more and has been injured a lot more.

Whether a product of regression, injury or not fitting into the system or culture of a club, most of the world's most expensive transfers have been flops. Money doesn't guarantee success.

The earlier calculation offers a simplified view of how to value players. Commercial considerations, off-field factors and the small matter of wages and bonuses must also be taken into account when evaluating footballers. But there certainly exists a fundamental disconnect between price and value in the market for players. Price (what clubs are paying) seems to have departed itself from value (what players are worth). Intelligent clubs can exploit this by recognising when a prospective buyer is willing to spend more on a player than his worth to their club. It takes a lot of courage to let a superstar leave, but it often represents a better deal for the selling club than the buying one.

Clubs generally tend to overpay for transfers, which naturally means that selling your best talent makes logical sense. Of course, this market dynamic makes purchasing players more difficult. When the entire market is inflated, how can you go about identifying undervalued prospects?

FINDING DIAMONDS IN THE ROUGH

When Manchester City bought Erling Haaland from Borussia Dortmund, their directors likely didn't spend too much time debating the signing. The number of potential centre-forward signings who could improve their squad could probably be counted on one hand. As one of the best and wealthiest teams in the world, the pool of players worth their time scouting isn't very large. The vast majority of clubs aren't in such a comfortable position. Bologna, for example, are the 100th best team in the world at the time of writing according to *FiveThirtyEight*'s global club rankings table, which uses xG data to rank the strength of hundreds of teams worldwide. There are an estimated 2,500 players globally whom Bologna could realistically target and that would improve their squad. Some of these players are employed by better teams, but many play for worse teams as well. Bologna not only have a budget many degrees smaller than Manchester City – which makes recruitment decisions more important and mistakes more costly – but they also have a much bigger pool of talent to sift through. When approaching the transfer market, each club is tasked with finding a needle in a haystack. For clubs like Bologna there are more needles, but also more hay.

Data is an incredibly useful asset for the Bolognas of this world. It allows clubs an instant overview of thousands of players, allowing them to cut through the overwhelming noise of the transfer window. Analytical recruitment isn't necessarily about finding the best player for your team. It's about filtering out the tens of thousands of worst ones. Ninety-nine per cent of the 2,500 players that would improve Bologna's team aren't even worth considering. They wouldn't suit the manager's requirements, wouldn't be able to fit seamlessly into their system, or wouldn't be the right signing for a host of other reasons. These players need filtering out, and data is the best filter. There are around 11,000 players residing in 450 teams spanning across 60 competitions who are capable of playing in the 'big 5' European leagues. Data can help narrow this view and find players who fit your system better than any scouting network ever could. Recruiting this way saves time and resources, and allows clubs to cast their net much wider than they'd otherwise be able to.

The recruitment process will tend to start with the manager, even in a smart, analytically minded club which seeks to distribute decision-making power between several stakeholders such as a Director of Football or Head of Recruitment. The head coach will give a detailed outlining of the kind of player they want to recruit for their team. The analysts will then filter through tens of thousands of prospects. The analytics department are tasked with weeding out as much of the hay as possible. In the manager's report of his ideal player, he will often include a list of key characteristics. Does the player need to have a high pass accuracy? How much xG does he accumulate per 90 minutes? Does he cover long distances during games?

That last question was of particular importance to Arsène Wenger when he was looking to replace Patrick Vieira back in 2004. The most important attribute that Wenger wanted in Vieira's replacement was the ability to run long distances. Statistical analysis back then was limited, but the distances a player covered in a match could still be accurately measured. Wenger consulted stats from various European leagues, before stumbling across a young midfielder by the name of Mathieu Flamini. The Frenchman, who was playing for Marseille at the time, was clocking an incredible eight miles per match. Wenger travelled to Marseille to see whether Flamini had the technical ability to match his dogged stamina. The Gunners boss was sufficiently impressed by what he saw. He signed Flamini on a free transfer and the player went on to achieve great success at Arsenal. Four years later, Flamini was sold to AC Milan in a deal nearing £10 million.

After analysts feed the manager's specifications into the club's software – players who run over 14km per match, players with a pass accuracy over 85 per cent, players with more than 0.40(xG) per 90 minutes, and so on – they are left with a much shorter list of potential targets. The metrics that are fed into the system can be repeatedly refined and made more specific, thus leaving fewer and fewer players who still match the criteria. Eventually, the analysts will be left with just a handful of players who still meet every condition. An overwhelming number of potential targets will have been whittled down to a more manageable group.

Once the data highlights potential needles, the recruitment team are employed to find out everything they can about the remaining shortlisted targets. Information on the player's injury history, wages, personal life and even their temperament will be assessed before any decision over signing the player is made. No stone is left unturned. Analysts will scour hundreds of hours of footage, providing insight into the strengths and weaknesses of each target, before they bring their reports forward to the manager, Director of Football, transfer committee or whatever decision-making entity the club deploys. The club will assess whether each target matches the initial brief the manager set. If a player gets the all-clear, the club will enter negotiations over a potential deal.

The transfer market is subject to the whims of supply and demand, the same as any other industry. Just as the number of consumers wanting the new iPhone pushes the price up, or the increased number of people who don't want to get wet in the rainy season inflates the cost of umbrellas, so too do the number of clubs shopping in a particular geographical market push the price of those players up. Ligue 1 serves as a good example. As the top league in a country renowned for producing talent, it attracts a plethora of scouts from around the world. Many players in this division are certainly good enough for other top leagues, but they're also likely to come at an inflated price due to heightened demand. In fact,

Premier League clubs have tended to pay 15–20 per cent more than 'market rate' for signings from Ligue 1 compared to players with similar statistical profiles who have heralded from other divisions.

Expected Goals models can help identify teams with similar-quality players who ply their trade in less fashionable leagues. As briefly mentioned earlier, *Five ThirtyEight's* club ranking table assigns a rating to hundreds of football clubs worldwide based on the xG they create and concede. The table ranks all professional European clubs (and many further afield) in order, as if they all played in one enormous league. We'll study later how to form such tables and why they're so useful in comparing clubs both within the same domestic division but also across leagues from different countries.

For now, consider Figure 7.3 which shows the rating (out of 100) assigned to each club in Ligue 1 and the Eredivisie. Each club is represented by a dot which is ranked in order of their quality within their league. Paris Saint-Germain are the left-most dot in the Ligue 1 line, reflecting their status as the best team in the division with a Power Rating of 82.6 at the time of writing. Ajax, the best team in the Eredivisie, aren't far behind them with a Power Rating of 80.0. On first glance it's immediately obvious that the overall quality of the French league is stronger. The average team rating in Ligue 1 is 61.6, compared to 52.8 in

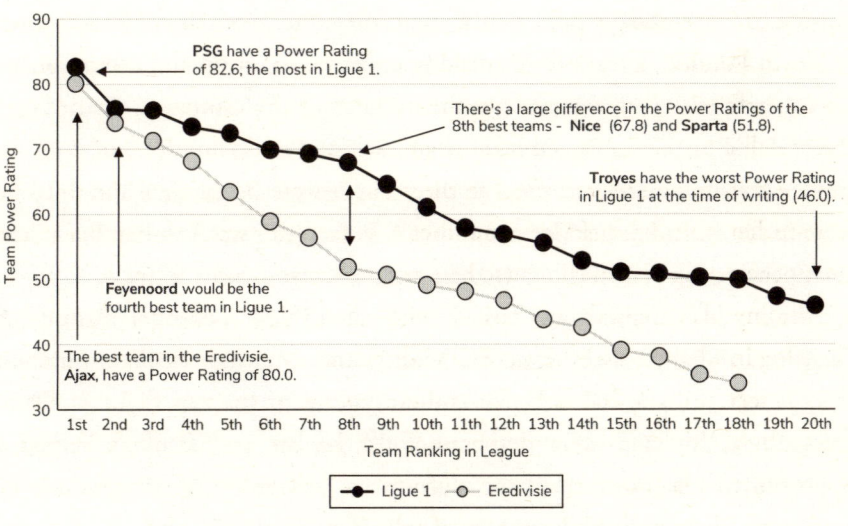

Figure 7.3: Team Power Ratings, Ligue 1 and Eredivisie

Eredivisie. However, the difference at the top ends of the table is minimal. Feyenoord, the second-best team in the Dutch league, rank higher than teams like Monaco and Lyon in Ligue 1. A smart club might spot this and realise Feyenoord's players, despite playing in the less fashionable Eredivisie, are actually better than some of the top players in France. The market is likely to undervalue these footballers given the lack of buyers shopping around in the Netherlands.

The fact that underrated players from less well-known leagues can be harvested at rock-bottom prices falls into the hands of Brighton and Brentford, who use chance-creation data from their owners' data consultancies to drive recruitment decisions. Remember that these companies collect bespoke xG data on almost every professional team and league throughout Europe. This allows them to cast their net much wider when approaching the transfer market and form valuable insight on players that might slip through the cracks for clubs with less vast data resources.

Greuther Fürth are a small German team who hail from northern Bavaria. In 2013/14 they finished third in the German second division, meaning they earned a two-game play-off against the team who finished 16th in the Bundesliga, Hamburg, for a shot at promotion. The tie was drawn 1-1 on aggregate but Greuther Fürth lost by courtesy of the away-goals rule. And that was it, a reasonably successful season for Greuther Fürth ended in failure and no one outside of Nuremberg would care. Except, they did. More than 500 miles away in North London, a team of Smartodds analysts had sat up and taken notice of Greuther Fürth's performances over the course of the campaign. According to their Global Justice Table, a model which rates and ranks hundreds of teams in a similar way to *FiveThirtyEight*, Greuther Fürth were better than the third-best team in the second tier of German football. In fact, this small club in Bavaria had been performing at a level higher than many *Premier League* teams.

Smartodds' analysts are tasked with identifying potential signings for Brentford and their sister team, FC Midtjylland. Matthew Benham purchased the Danish club in 2014. Broke and struggling in the top flight of Danish football, Midtjylland were desperate for a backer and Benham sensed an opportunity. He transformed the club into a laboratory for his revolutionary analytical ideas and philosophies. Midtjylland's players and coaches were sceptical of Benham's statistical methods, but they were desperate for investment and realised that beggars can't be choosers. Benham had only just taken over the Danish club when his models flagged Greuther Fürth as a club massively outperforming expectations. The small German side were creating a huge amount of xG while conceding very little at the other end. No one else in European football realised it, but this squad likely contained a treasure trove of undervalued talent.

Benham and his analysts took a deeper look into Greuther Fürth's playing staff and came across a commanding central defensive midfielder called Tim Sparv who had played more minutes than any other player that season. Smartodds' analysts identified his presence as integral to Greuther Fürth's exceptional performances. Midtjylland spent a measly £255k to secure the services of Sparv, who became the first major 'analytics signing' of the system that would end up making Midtjylland and Brentford masters of the transfer market. In Sparv's debut season at the club, Midtjylland won the Danish Superliga for the first time in their history. At the start of the next season they beat Southampton, a club of much greater size and wealth, over two legs to qualify for the group stage of the Europa League for the first time. They progressed through to the knock-out stage, where they were drawn against Manchester United – a club that had completed the treble in the year that the Danish side were formed, 1999. Midtjylland pulled off a famous upset to defeat Louis van Gaal's team in the first leg of the tie (although were comfortably beaten at Old Trafford in the second leg).

Midtjylland continued to supplement their squad with Sparv-like signings: players performing for teams who the wider footballing world perceived as average or poor, but who in fact were ranking extremely highly on Smartodds' Global Justice Table.[9] Finding the undervalued teams leads you to the undervalued players. Benham understood that, as with gambling, the recruitment of players is about spotting the value in the market. Each player is a bet which you hope

---

[9] The company use the term 'justice' as a nod to the fact that the table shows where each team *deserves* to be based on their performances. Again, we'll study these ranking systems in more detail later on.

pays off in the long run. Midtjylland's methods proved successful and the club won the league title three times in Sparv's six-season spell.

The Global Justice Table not only shows the overall ranking of a club, but also the attacking and defensive strength of each team. If one of Benham's clubs needs a new defender, his analysts can look for teams who are not just positioned highly on the global xG ranking system, but who are also conceding very few scoring opportunities. A team who concedes a small number of high-quality shots are more likely to have a strong defensive line, meaning their defenders are worth a closer look. On the other hand, a team who create a high number of dangerous attacks are likely to have a strong set of forward players.

The bottom end of the Premier League and the top end of the Championship doesn't differ all that much in terms of quality.

Benham's model highlights an interesting truth about domestic league systems. It's generally assumed that teams in the top tier of a country's football pyramid must all be better than the teams in the league below. After all, promotion and relegation should ensure the good teams and bad teams are fairly sorted according to their ability. In fact, the Global Justice Table shows this isn't necessarily the case. There isn't so much a *gap* in quality; there's an *overlap*. The success of recently promoted Premier League teams such as Wolves, Leeds, Fulham, and Brentford speaks to the difference in quality between the top two tiers, as does the fact that relegated teams often struggle to bounce straight back up. The bottom end of the Premier League and top end of the Championship doesn't differ all that much in terms of quality.

If the standard of football is comparable in the league below then sending players out on loan to these teams becomes more viable. The match day opposition is perhaps more capable than you initially thought, as is the standard during training sessions. Recruitment is the other obvious angle. The market may see certain teams and players as 'second-tier' quality, but if a lot of second-tier clubs are actually already performing at the level required for the top flight then there are likely unappreciated players hidden in these leagues. Research suggests the overlap becomes even greater further down league pyramids, between the second tiers and third tiers, third tiers and fourth tiers, and so on.

Moreover, talent from lower down the league system often has a greater ceiling. Consider two young track runners, both specialising in 100-metre sprints. One has been conditioned in this format his entire life, spending his youth at a state-of-the-art training facility and being tutored by the best coaches in the sport. He can run 100 metres in 10.5 seconds. The other runner is rawer, hailing from a poorer corner of the world and has mainly taught himself the art of sprinting. His time clocks in at 10.8 seconds. If you were looking to invest time and resources into one of these runners, which would you choose? The raw 10.8 probably has more potential than the trained 10.5. Likewise, footballers from lower down the league system can probably improve more than players of an equal ability who have spent their careers in the upper echelons. The English pyramid possesses countless non-league divisions that reside under the four professional leagues. Recent examples of Charlie Austin, Jamie Vardy, Jarrod Bowen, Michail Antonio, Chris Smalling and numerous others suggests there are plenty of non-league players who would be comfortable competing in the top tiers. Teams and players aren't necessarily fairly distributed according to quality: it's more fluid than we are led to believe by the structure of the 19th century league system format.

It's not just the analytically driven teams who have appreciated the value in signing players from untapped markets. Ange Postecoglou used his knowledge and experience of the J1 League, the top tier of Japanese football, to harvest a wealth of talent when he was appointed Celtic manager in 2021. In his first 18 months at the Scottish team, he signed no fewer than six Japanese players on to the books. Postecoglou knew the J1 League was comparable to many of the top divisions in European football and the players were capable of making the move. The perception of value (or lack thereof) in Japanese football is a massive inefficiency in the European transfer market – one Celtic have had no shame in exploiting. Celtic's expenditure on the six J1 League players they signed was around £10 million, but leading models put the cohort's value at £35 million or so at the time of writing. Postecoglou's left-field recruitment strategy preceded a World Cup campaign where Japan beat Germany and Spain and were knocked out by semi-finalists Croatia on penalties. Kyogo Furuhashi ended the 2022/23 season as the top scorer in Scotland with 27 goals and has been linked to some of the top clubs in the world. The market for Japanese players certainly doesn't seem likely to remain unappreciated for too much longer.

SMARTER RECRUITING

Recruiting undervalued players from foreign or less fashionable leagues can clearly pay dividends. But it's important to ensure new recruits fit the brand of football your team plays. When it comes to new signings, style is equally as important as substance. The very first step of smart recruitment is to define the system you want to play and work out which areas of that system you need to strengthen. The manager will often handle this stage of the process, building a profile of the type of player he wants to sign and presenting that profile to the Head of Recruitment or whoever manages the player-acquisition process. The recruitment department will ask the manager about the player and role they're looking for. 'Tell us your ideal footballer for this role. Describe what things they do. What qualities do they absolutely need, and what are the 'nice-to-haves'?'

A manager might want a striker who links play well, runs the channels and carries a goalscoring threat. Perhaps the club can only afford a couple of these things – players who excel in many remits will often carry a heftier price tag. Can you sacrifice one of these qualities and reproduce it elsewhere on the field? Billy Beane always spoke about replicating a player's output 'in the average.' A striker who is sold will have made certain contributions to the team that need replacing. The club can look at players in comparable leagues who produce a similar statistical output. They can match their profiles and ensure the new striker will fit in seamlessly.

Prioritising the collective over the individual is crucial. If a striker doesn't link up the play, can you compensate by signing or promoting a more creative midfielder? Football is a game of yin and yang where the balance of a team must be close to perfect for them to succeed. Yes, teams sign individual players. But these players are cogs that fit into the wider machine. You can buy the shiniest, most expensive cog in the world, but if it doesn't fit in with the wider mechanism then it will be rendered next to useless. If you're a pressing team, like Liverpool under Jürgen Klopp or Leeds under Marcelo Bielsa, then you need to ensure new signings are able to perform in a high-intensity, physically demanding system. If you're a team looking to control possession, like Pep Guardiola's Manchester City or Vicente del Bosque's Spain, then you need ball-playing centre-backs and technically skilful midfielders. Teams are becoming better at modelling the stylistic fit of players and making new signings accordingly.

Clubs are also increasingly aware of the importance of player trading in securing profits. Buying young players, developing them, and selling at a higher price point is key to operating sustainably. 'Resale value' has become a slogan for teams looking to exploit the market for revenue-driving purposes. However, as clubs become progressively more mindful of signing young players, the age of footballers is becoming increasingly factored into initial pricing. In financial markets, information is usually 'priced in.' For example, the value of Pound Sterling doesn't simply react to the outcome of events such as general elections and pandemics, but moves in anticipation of them. Everyone thought that the Pound would drop if a no-deal Brexit was agreed, so when the probability of that outcome increased before the event had taken place, the Pound dropped accordingly. The same principle applies to the evaluation of footballers. Paris Saint-Germain bought Kylian Mbappé for two reasons. Firstly, he's really good at football. Secondly, he was likely to get even better at football and had many years of output ahead of him. Both these reasons were factored into the €180 million fee the club paid Monaco in 2018. If everyone thinks that a player's value will rise, the market should reach a state where there's roughly a 50-50 split of the buying club actually realising a profit on the deal. It's dangerous to assume all young players will increase in value simply because they are young. What clubs should be looking out for is *undervalued* talent – the players other clubs don't think will increase in value, but who actually will. The key is to find the misshapen fruit.

Figure 7.4 shows the general transfer value of players at different stages of their career. A player is roughly twice as valuable as a 20-year-old than as a 26-year-old, and roughly five times as valuable at 20 than at 30. The older generation are likely beyond their peak performance levels and will command less resale value, which is factored into their pricing. Every year, a player's value

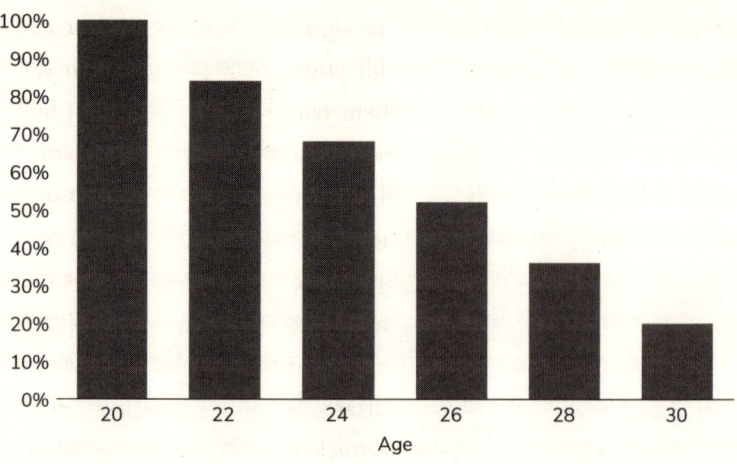

Figure 7.4: Player Transfer Value

drops by about 8 per cent. This reality is worth remembering for clubs looking to secure profits in the transfer market.

Identifying and buying young talent is one thing, but it's important for teams to actually make space for hot prospects in the first-team squad. Take Chelsea as an example. The club's recruitment operations enabled them to move first on players like Kevin De Bruyne, Mohamed Salah and Romelu Lukaku. But the impressive feat of signing such talent early on was undermined by the philosophy of the club in actually giving these players minutes. The curse of wealth means you can have too many superstars; players in their peak years on massive wages who the manager deems it impossible to drop. This stifles the pipeline of young talent and hinders the profitability of clubs looking to achieve sustainable transfer-market practice. All three of those Chelsea youngsters went on to find great success at rival teams.

Borussia Dortmund are a club on the other end of the spectrum to Chelsea. The German outfit have become known for attracting young talent. They've built a reputation for giving the next generation game time, developing them and allowing them to move on when the time comes. All three of these steps are selling points for young players deciding where to ply their trade. Dortmund have historically provided a meaningful role within the squad for youngsters, striking the right balance between playing youth alongside those in their peak years.

The decision of whether or not to sign a player can cost or earn a club millions. As such, it's important to fill your talent-identification department with the brightest brains and equip them with the best tools. But how can we evaluate the performance of our scouts, data analysts and other members of the recruitment team? Each employee will submit hundreds of reports on potential signings each year, sometimes recommending them and sometimes condemning them. It's fairly easy to track the progress of new recruits who come through the system, but most of the players (recommended or otherwise) will never actually sign for the club. Clever clubs will keep a record of the players their club *didn't* sign, as well as the ones they did. Starlizard, Smartodds and consultancies such as Twenty First Group can accurately track the performance level of a player throughout their career. Basic indicators such as the xG output of the player's team, as well as the number of minutes he is playing, can be cross-referenced with more nuanced individual data analysis to provide a rating for the player which benchmarks him against his peers.

Suppose three separate analysts all recommend their club signs a different footballer of similar perceived ability: Player A, Player B and Player C. The club finds the first analyst's report the most convincing, so signs Player A. Usually, the club (and external onlookers) would assess the performance of its recruitment department on the success of this player. But tracking the performance arcs of all three players will give a more profound view of which analyst made the best recommendation. In the theoretical example presented in Figure 7.5, the analyst who suggested Player B actually made the best assessment. Repeating this process for each of the hundreds of players each analyst or scout puts forward will give a better view of which members of the recruitment team are consistently providing the best recommendations. All organisations should develop effective measures for tracking the performance of their employees. Just as clubs use data to measure the output of their playing staff, so too can they use it to evaluate the performance of their non-playing staff. Such 'performance appraisals' of analysts and scouts help improve talent identification.

There is a clearly defined food chain which exists within football. At the top are the clubs with seemingly bottomless pockets. These sides will almost always finish at the top end of the league table; anything less than the best is failure.

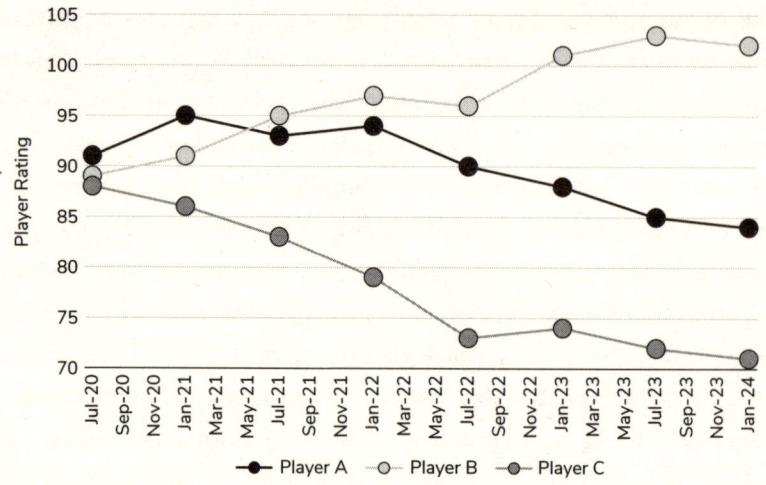

Figure 7.5: Player Rating Tracker

There is no need for these clubs to innovate in the transfer market. The 'super clubs' can use their financial brawn to dominate the world of football, but the overwhelming majority of teams are not in quite such a luxurious position. Ninety-nine per cent of professional clubs could be labelled as both 'predators' and 'prey' in the football food chain. These sides are in a constant battle to try and increase their affluence. They all have teams who are richer than them and who will swoop for their best players. Similarly, they have teams who are less well off than themselves and whom they can lure players away from. Footballers are drawn to money like moths to a flame and the best players will naturally gravitate towards the richest clubs.

This makes the importance of transfer-market success two-fold. Firstly, getting recruitment decisions right is the primary way to increase your own wealth and work your way up the food chain. Teams like Brighton, Brentford, Royale Union Saint-Gilloise and Midtjylland have used player-trading to continuously upgrade the ability and value of their squads, rocketing themselves up the league pyramids in the process. Secondly, because footballers follow the money, becoming better at spotting undervalued players is the only way to attract high-calibre talent to your team. All other things being equal, the richer clubs will triumph. Those on restricted budgets have to become *smarter*. This means embracing the role of analytics, innovating, and staying ahead of the curve. They need to appreciate the doctrines outlined in this chapter and become better at assessing the true quality of players. Let the competition buy the good-looking fruit, while you enjoy the tasty misshapen goods they've thrown away.

CHAPTER SUMMARY

- Player trading is perhaps where application of analytics is most crucial – transfers are incredibly important to a team's success but also incredibly difficult to get right.
- Recruiting effectively is one of the fastest and most efficient ways to increase your own club's wealth and buying potential.
- Good players tend to be intrinsically overvalued in the market for a host of reasons. Smart clubs can cash in on this.
- Forward-thinking teams are using a range of analytical tools to help them identify undervalued talent and make smarter recruitment decisions.
- A player's value drops by roughly 8 per cent year-on-year, meaning it is critical that players are recruited at the right age if they are to still have a high transfer potential.
- Accurately assessing player quality is key to understanding the science of winning football matches.

8

THE TACTICS BOARD

HOW DATA INFLUENCES WHAT HAPPENS ON THE PITCH

*'The ball is round, the game lasts 90 minutes,
and everything else is just theory'*

Sepp Herberger, West Germany manager (1950–1964)

The Euro 2020 final was famously late. The tournament was delayed a year to allow domestic club competitions to be completed as football authorities grappled with the unprecedented carnage caused by the coronavirus pandemic. UEFA decided to maintain the name 'Euro 2020' despite the postponement to 2021, supposedly because they wanted to 'keep the vision' for the competition. In reality, the decision actually boiled down to a matter of cost – changing the name would have required an expensive rebranding campaign as many of the tie-in products and logos had already been designed. The keyrings and t-shirts had been manufactured for a tournament called 'Euro 2020'. The competition's final was late on another count: the extra time and penalties it took to decide the contest meant it was nearly 11pm at Wembley Stadium by the time Gianluigi Donnarumma saved Bukayo Saka's penalty to win it for the Italians. The result left many people wondering, 'In a game that was famously late, did England score too early?'

Football analytics is littered with unfamiliar and confusing terminology: Expected Goals, Passes Per Defensive Action (PPDA), half-spaces and so on. Among these, one refreshingly simply named metric presents itself. 'Game state' refers to whether a team is winning, drawing, or losing when a certain action or set of actions takes place. This can provide crucial context to match data and explain the tactical choices implemented by either side. For example, a losing team might be inclined to take a greater number of shots and throw players forward in search of an equaliser, while a winning team might be prone to sitting deep, containing the opposition and prioritising defensive resilience over chance-creation.

England deployed the latter tactic to its full extent during the Euro 2020 final. Luke Shaw scored less than two minutes into the match, half-volleying Kieran Trippier's cross home to put England ahead. The shot was worth 0.35(xG), indicating the average player would be expected to score this chance about one in every three occasions.[10] The attempt is represented by the almost immediate uptick in England's cumulative xG trendline shown in Figure 8.1. This chart depicts the rolling Expected Goals total for both England and Italy over the course of the final. As you can see, England created most of their xG in the first two minutes of the two-hour contest.

England accumulated 0.60(xG) overall in the final, which pales in comparison to Italy's 2.23(xG). Without any other context, we might conclude Italy were by far the stronger team and utterly dominated England. However, game state provides an insight into the strategic approach of both teams. England's game state between the 2nd and 67th minute was +1. In other words,

---

[10] Generally speaking, when fans are shown clips of shots and asked to guess the xG, they tend to overestimate the figure. In other words, goals are harder to score than we think. You'd be forgiven for thinking Luke Shaw's effort merits a larger xG than 0.35, but controlling a half volley of this nature from an unideal angle with two opposition players in the way isn't as easy as it might seem.

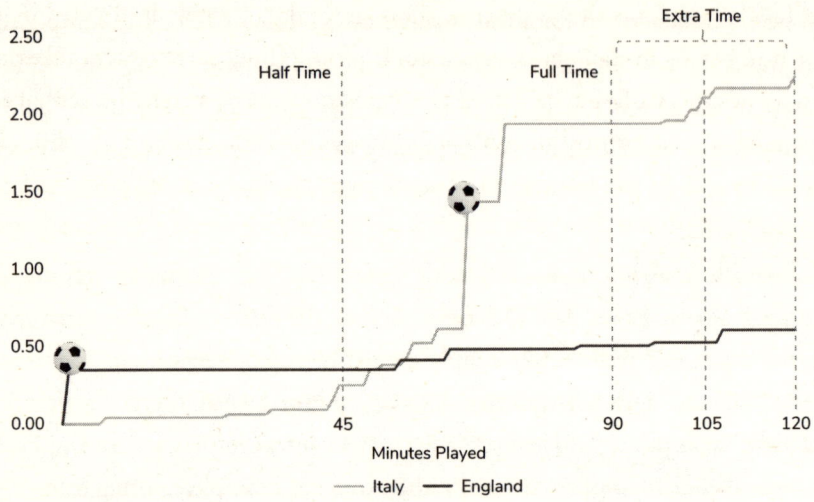

Figure 8.1: xG Timeline – Euro 2020 Final Italy (2.23) 1-1 (0.60) England

they were winning by one goal. England's tactical choice from here on in was to sit deep and attempt to protect their lead. The timeline shows they mustered only 0.13(xG) over the 65-minute period where they were leading. Meanwhile the Italians became increasingly attacking and managed to create a handful of chances, particularly after half time. Italy broke through with just over 20 minutes of normal time left and England were unable to revert to an attacking mentality. The game state gives us more context into how the match played out rather than simply looking at the xG scoreline. If England hadn't scored so early, they would have probably created more xG during the match.

We can study the impact of leading, drawing or trailing on the attacking nature of teams by looking at shooting patterns during various game states. Football analysts have long since recognised the correlation between the number of shots a team takes and their proclivity for winning matches. A key metric for proving this hypothesis is Total Shots Rate (TSR). TSR follows a simple formula but provides a decent indicator of a team's performance. The equation is as follows:

Total Shots Rate = Shots For / (Shots For + Shots Against)

The TSR of a team is the number of shots they take in a match, divided by the total number of overall shots which took place. Thus, a team who takes any number of shots without reply from the opposition can be assigned a TSR of 1.0. A team who has six shots but concedes four attempts will carry a TSR of 0.6 (six divided by 10). If the number of shots of both teams are equal, then they will both hold a TSR of 0.5. Here is the TSR of both sides if we plug in the data from the Euro 2020 final:

Italy had 20 and England had 6 shots,

Italy's TSR = 20 / (20 + 6) = 20 / 26 = 0.77

England's TSR = 6 / (20 + 6) = 6 / 26 = 0.23

Italy's Total Shots Rate was 0.77, while England's was just 0.23. The TSR of both sides is always 1.0 when added together and the average of the two TSR totals is always 0.5. This is because every time a shot is taken by one team, the other team concedes a shot. There is perfect parity between the two figures.

There is a strong correlation between a team's average Total Shots Rate and their position in the league. This is intuitive: the best teams take the most shots. But the problem with the formula is that it treats all shots equally. In football, some chances are going to be a lot easier to score from than others. An effort from two yards out has a higher probability of hitting the back of the net than a shot from 40 yards out. TSR doesn't take into account the quality of shots, only the quantity. The metric would work perfectly if all teams created an equal number of good chances as poor chances, but it's unrealistic to assume the distribution of low-quality shots to high-quality shots is equal for all sides. Arsenal, for example, gained a reputation under Arsène Wenger for passing the ball to death. The Gunners were often accused of 'over-playing' and were encouraged by their fans to take more shots. While Arsenal's lack of attempts at goal may not have seen them rank too favourably on the TSR scale, the shots they did take tended to be of high quality. Arsène Wenger was a known 'xG believer,' so it's no surprise he was ahead of his time in prioritising the creation of big chances over long shots.

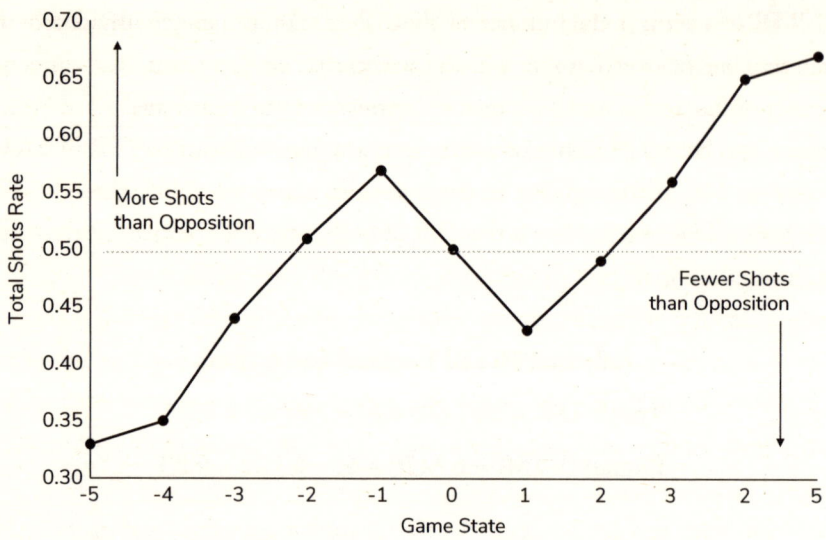

Figure 8.2: TSR by Game State, Selected European Leagues 2014–2022

Back to game state. Figure 8.2 shows the Total Shots Rate for each game state across a large number of European leagues over nearly a decade. Teams with a game state of '0' have a TSR of 0.50. In other words, sides who are drawing take on average just as many shots as their opponents. What's interesting is what happens either side of this equilibrium. Teams who lead by one goal take fewer shots than their opponents (0.43 TSR), while teams who are trailing by a goal take more shots (0.57). The same is marginally true for teams who are winning or losing by two goals (0.49 and 0.51 respectively). Teams winning by three goals or more tend to heavily outshoot their opponents, presumably because the gulf in class is large enough for them to fully dominate the match.

These findings have a material impact on the nature of how football matches are played. A closely trailing team generally take more shots than the team who lead them. This is surprising. You would assume a team who is winning would likely be better than their opponents and create more shots as a result. The fact that 57 per cent of shots taken when there's one goal difference between two sides is attributable to the losing team shows the extent to which sides sit deep and look to protect leads. England's defensive approach against Italy is a typical example of how teams play when they're winning.

Bill James, the father of baseball's analytics movement, commented that 'there is a negative momentum in the world that acts to reduce the difference between good teams and bad teams, between strong teams and weak teams'. James noted that the balance of strategies always favours the team that is behind in a game, because the other team becomes defensive, is less likely to take risks, and probably experiences nerves which are detrimental to performance. Psychology tends to pull the winners down and push the losers upwards. This momentum also exists on a macro level: poor clubs take more risks and are more open to innovation, whereas rich clubs can become stagnant because they always have the option of falling back on their vast wealth. Innovation tends to happen with the Brightons and Brentfords of this world, not with the Manchester Uniteds or Tottenhams.

Game state can be applied over the course of a season to explain various trends and provide context to teams' attacking and defensive outputs. Under José Mourinho's stewardship, Tottenham were famed for shutting up shop as soon as they took the lead. Figure 8.3 shows the Expected Goals per 90 minutes (xG/90) that each of the 'big 6' teams created when in a game state of +1 during the 2020/21 season. Chelsea carved out the most opportunities when leading by a

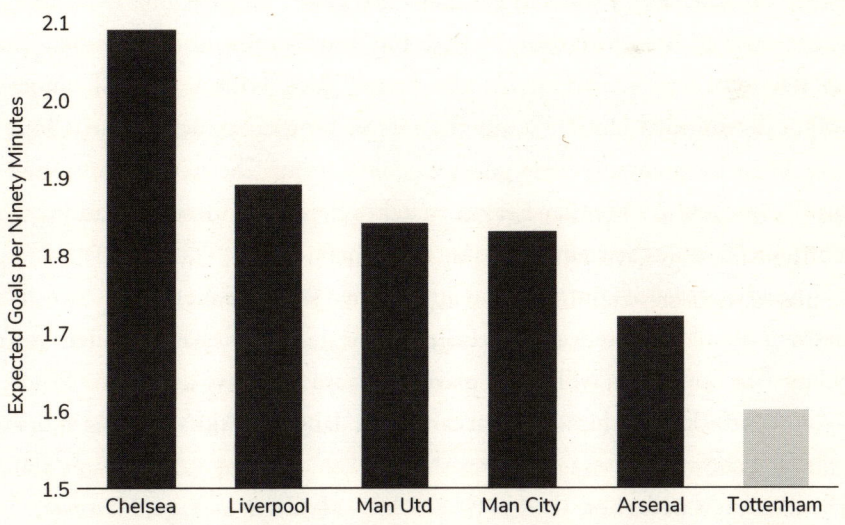

Figure 8.3: Chance Creation When Winning by One Goal, Premier League 2020/21

goal, looking to put the game to bed. Liverpool, Manchester United and Manchester City took a more balanced approach, still creating chances but also being wary of protecting their advantage. Jose Mourinho's Spurs achieved the lowest attacking output of this cohort when leading by a goal, creating 1.60(xG/90) during this particular game state. Tottenham fans grew frustrated by watching their team retreat every time they took the lead, and perhaps they were within their rights to feel this way because Spurs also conceded more chances than any of the other 'big 6' teams except Manchester United when leading by a goal – allowing 1.41(xG/90).

Game state shows us that trailing teams tend to outshoot leading teams, but it doesn't necessarily tell us that this is the *correct* way to play football. Elite clubs are often able to sustain similar shot levels at any game state. Moreover, studies have shown it's almost always better to continue taking risks and go for a second goal than to rest on your laurels. Brentford carried out an analysis which revealed the merits of gunning for a second, match-clinching goal. The Bees adopted this approach in the Championship and promptly became known for blowing teams away, finishing as league top scorers in 2019/20 and 2020/21. Phil Giles said, 'One of the owner's big bugbears is when the team are 2-1 up in the final few minutes and everyone drops off. He's like, "I want to keep on attacking, I want to get that third goal".' In their first two Premier League seasons, Brentford took 39 leads but only lost one of these games. That was the joint-best record across the league (tied with Liverpool, another analytically minded club). Could the reason be their reluctance to adopt a more defensive mentality after going a goal in front? The stats seem to say so. Out of their first 19 Premier League matches Brentford took the lead in, they went on to double their advantage in 16 of them.

The advantages of continuing to attack when you take the lead are two-fold. Firstly, it should theoretically be easier to create better chances when you're leading. The opposition will throw players forward and leave themselves exposed to counterattacks, which means you can create dangerous situations for yourself. Figure 8.4 shows the average xG per shot broken down by various game states. When a team is tied, shots tend to be worth 0.11(xG) on average. However, that figure jumps to 0.14(xG) when a team leads by a goal. By the time a team is two

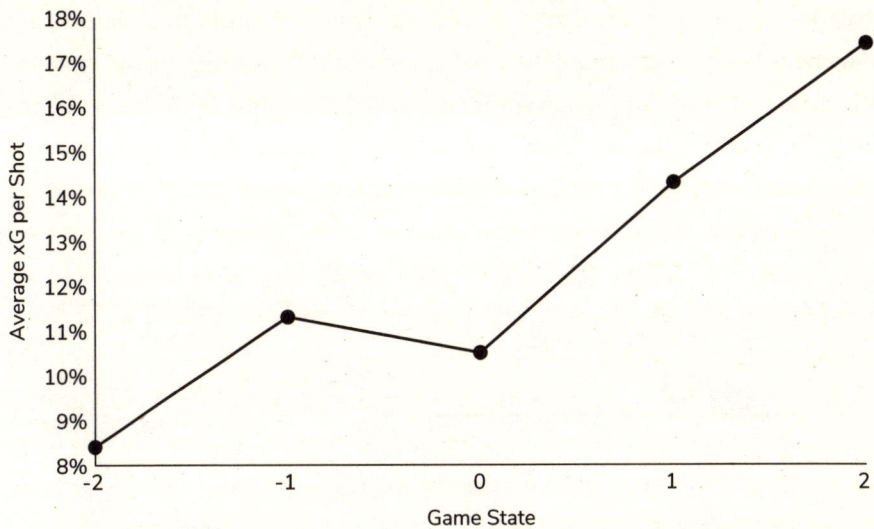

Figure 8.4: xG per Shot by Game State, Selected European Leagues 2014–2022

goals ahead, they create shots worth an average of 0.17(xG). Less defensive pressure on the shooter opens the door for higher-quality chances.

This data informed Brentford's tactical decision-making. They realised they could have the best of both worlds: defend deeper when winning, encouraging opponents to play expansively and push players forward, but still throw their own players forward when counterattacking in an attempt to find that killer goal. Of the 17 clubs to play in both the 2021/22 and 2022/23 Premier League campaigns, the Bees were the most direct when in a winning position. They averaged both the fewest passes per sequence and the fastest direct up-field speed when leading. Brentford's threat in transition, varying between flowing passing moves, direct dribbles and quickly launched balls forward, allowed them to spring up the field and put the game out of reach of opponents. Their big-chance conversion rate when winning was around 60 per cent, the highest of any team to play in either of those Premier League seasons. Brentford, being the xG-guided team they are, deal largely in 'big chances'.

The second advantage of continuing to press for a second goal: it prevents unwanted pressure mounting on your back line. The old adage dictates that 'attack is the best form of defence.' Even in the Euro 2020 final, England

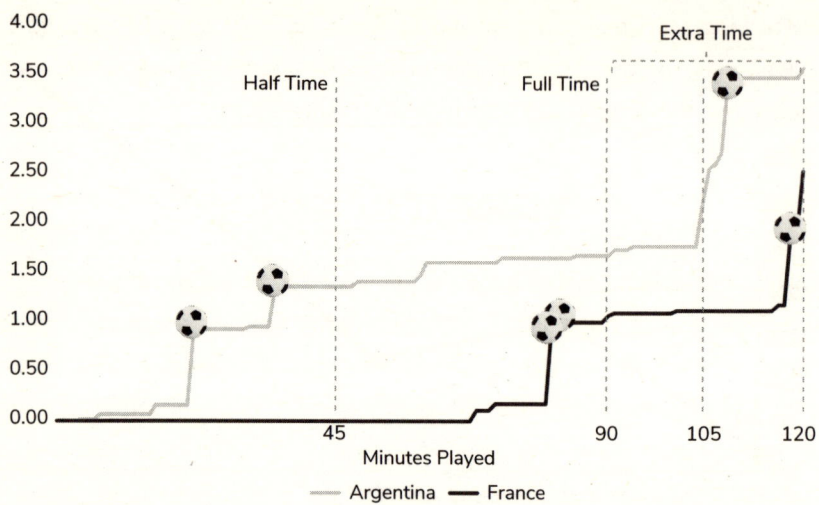

Figure 8.5: xG Timeline – 2022 World Cup Final, Argentina (3.57) 3-3 (2.53) France

conceded more xG (1.43) when their game state was +1 than when it was 0 (0.80), despite both game states existing for a similar amount of time. This is obviously just a one-match sample, but what would have happened if England had continued to play on the front foot and looked to create chances rather than dropping off? Perhaps they'd have had a better chance at winning their first major trophy since 1966.

Argentina fell victim to the same mistake in the 2022 World Cup Final. They dominated the early exchanges and were two goals ahead by half time. France weren't allowed a single shot until the 68th minute. But once Argentina sat deep in an attempt to see the game out, they invited France on to them and Kylian Mbappé worked his magic. The saying goes that '2-0 is a dangerous lead' – and so it was for Argentina. They created 1.36(xG) in the first 36 minutes to give themselves a game state of +2. However, during the 44 minutes they held this two-goal lead they only created 0.29(xG). If they'd have kept their foot on the gas and kept attacking with the same intensity, they might have scored a third goal and put the game to bed. Instead, they curbed their free-flowing attacking nature in favour of a defensive approach. With the game once again tied, Argentina created a whopping 1.83(xG) between the 82nd and 108th minutes of the match. They certainly showed a better ability to switch between defensive and attacking gears than England did in the Euro 2020 final. Of course, it all ended happily for the Argentinians on penalties, but there are lessons to be learned about the dangers of retreating once taking a lead.

WINNING AT ALL COSTS

A simple truth guides how we should approach the various game states (winning, drawing and losing), but it's one we often forget: *wins are three times as valuable as draws*. In 1981, the governing bodies of football decided to encourage more attacking play by awarding three points for a league win rather than two. Ever since, a victory has been twice as good compared to a draw (difference of plus two points) than a loss is as bad compared to a draw (minus one point). So, when at a game state of '0', it's almost always worthwhile taking risks and attempting to win the match.

Suppose Smart FC are drawing a game with 10 minutes left on the clock. The manager, Pep Smartiola, has a decision to make: do we hold on for the one point or attempt to push forward and grab all three? Smartiola consults the team's analysts who estimate a cautious approach, whereby they bring on defensive players and sit in a low block, meaning they'll almost certainly see the game out for a draw. If they decide to be more risk-taking, throwing on offensive players and committing numbers forward, they'll increase their chance of grabbing a winner but also leave themselves exposed at the back. They could also take a neutral approach somewhere between these two extremes. Which tactic should Smartiola deploy?

Figure 8.6 uses some simple Expected Value equations to help answer this question. Taking the cautious approach would mean a 5 per cent chance of getting three points (meaning the strategy is worth 0.15 xPoints), a 90 per cent of getting a point (0.90 xPoints) and a 5 per cent chance of getting no points (0.00 xPoints). This means if the match was simulated a large number of times, the team would expect to collect 1.05 points in the long term. The same calculation reveals the neutral approach to be slightly better (1.15) and the risk-taking approach to be the best (1.30). Even if we assume the risk-taking attitude means Smart FC are more likely to concede than score, the approach

	Cautious	Neutral	Risk-Taking
Win Probability	5%	15%	30%
Draw Probability	90%	70%	40%
Loss Probability	5%	15%	30%
Win xPoints	0.15	0.45	0.90
Draw xPoints	0.90	0.70	0.40
Loss xPoints	0.00	0.00	0.00
Total xPoints	1.05	1.15	1.30

Figure 8.6: Game Strategies When Drawing With 10 Minutes Left

still comes out very well. Suppose the analysts are incredibly conservative and calculate that throwing players forward means they'll have a 25 per cent chance of winning, a 40 per cent chance of drawing and 35 per cent chance of losing, Smart FC will still have the same projected points as if they take the neutral approach (1.15).

When drawing, you're better off taking risks and pushing for a winner. Even though you might end up losing, it's worth it. The league standings are often sorted by the number of wins a team has. Look at the current table of the club you support; you'll probably be surprised at how well it's filtered by the number of games each team has won. There might be a couple of instances where sides have drawn lots of games and be positioned higher than certain other sides who have won more, but generally speaking the table will be sorted by wins. Draws are largely inconsequential and overvalued.

This logical stance goes against human nature, which is naturally loss averse. Loss aversion refers to our tendency to place more value on avoiding losses than on acquiring gains. This means people feel more pain when losing something compared to the pleasure they experience when gaining something of equal value. Loss aversion is a fundamental part of human decision-making

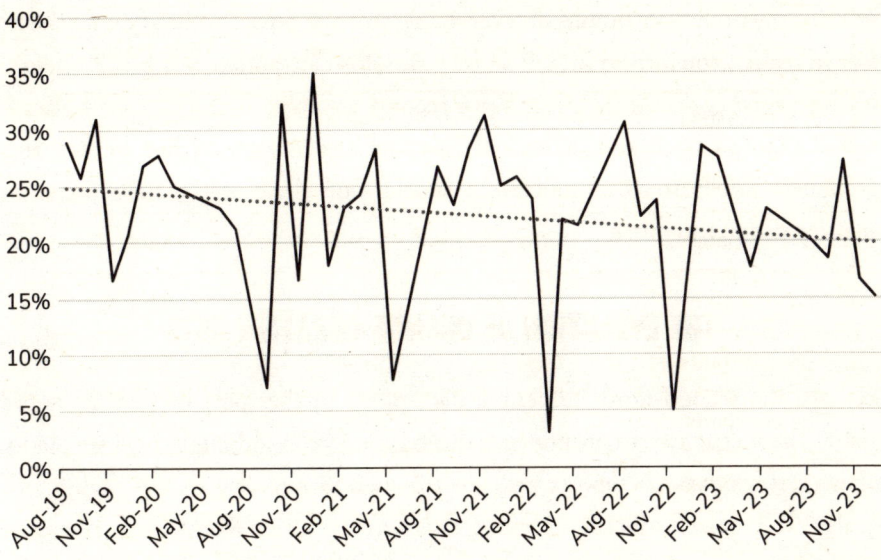

Figure 8.7: Premier League Draws, 2019–2023

and has been observed across a wide range of contexts. For example, people tend to hold on to losing stocks for too long, hoping that the market will turn in their favour. They fear the pain of realising a loss and often continue to hold on to investments even when they continue to lose value. Another example is the reluctance of people to give up something they already possess, even if it's not of much value to them. We cling to old possessions like clothes or toys, even if we don't have a practical need for them. Loss aversion is a powerful cognitive bias that shapes our decision-making in ways that are not always rational. The way league points are assigned in football means we must struggle against the natural gravitation of loss aversion, because the optimal strategy is to risk losing in order to seek wins.

Despite our natural loss aversion, clubs appear to be cottoning on to the reality that draws really aren't that helpful. The draw rate in the Premier League has declined significantly over the last few seasons as teams appear to be crunching the numbers and working out the benefit of taking risks in search of all three points. Some might use this as evidence to reinforce the notion a team should sit deep and look to maintain their lead when going ahead. If wins are so crucial, surely you should do everything in your power to hold on to them? But in reality, the importance of victory shouldn't invalidate our earlier theory: it's better to continue to push for a second goal when you're a goal ahead, rather than sit back and defend. If England and Argentina had heeded this guidance, the Euro 2020 and 2022 World Cup finals might have taken an entirely different course.

THE EVOLUTION OF CHANCE-CREATION DATA

Perhaps the second-most influential data point in the sport of football is the assist. The metric rose to prominence during the 1990s, although there are plenty of examples of match reports prior to this period referring to the 'making' of goals. FIFA started officially tracking assists at the 1994 World Cup. The assist reached such acclaim over the next few decades that the official Playmaker of the Season award was introduced in 2017/18 for the Premier League player with the most assists.

A large driver of the assist's rise to fame has been the success of fantasy football. Fantasy games have intrinsic links to the popularisation of data across a number of sports. In baseball, many of the early adopters of Bill James' sabermetrics revolution were fantasy baseball players who wanted to select better players for their team. Voros McCracken, a bored paralegal based in Chicago, virtually rewrote the rulebook on how to judge the ability of pitchers when doing research into whether to draft Chad Bradford for his fantasy team. He realised pitchers had no control of whether the ball fell for a hit once it got put into play – a notion that seems obvious now but was revolutionary at the time he made the discovery. What was previously regarded as the pitcher's responsibility was simply luck.

Fantasy football is a breeding ground for budding football analysts. There is a noticeable crossover in interests and skillsets between those who play games like Fantasy Premier League and those who have embraced xG. Indeed, many fantasy players use stats like xG to help guide their strategic decision-making. Should they draft in a certain in-form striker? Well, the xG data shows he's actually been scoring a lot of long shots and his scoring rate is likely unsustainable. How about that unfashionable, unpopular defender? Well, it seems his team are actually conceding very little xG so he's likely to keep more clean sheets in the future. Fantasy football is random, almost as random as football itself. Those who play it can do a great deal of research and gain a strong grasp on which players are worth selecting, but at the end of the day Lady Luck will have a big say in who succeeds. The game treads the line between luck and skill in a similar way to the sport it mirrors, which keeps its users engaged and lays down a challenge to those who wish to understand the truth of what brings success in football. Perhaps that's why FPL now attracts more than 10 million users per season. Assists are a key player scoring metric in any fantasy game worth its salt, so increased interest in measuring how and why they come about has coincided with the growing popularity of these games.

There are a lot of problems with the assist metric and how it assigns supposed value to players who accumulate them.

The assist is a funny stat. It lies somewhere in the middle ground between relevant and irrelevant. Some people like it – after all, it does offer a description of an event which actually took place. Detractors of Expected Goals will attempt to invalidate the metric by proclaiming, 'xG doesn't actually happen. It's an abstract, speculative stat. I only care about *actual* goals.' For these xG non-believers, at least assists are tangible actions that certifiably take place. They offer a description of something which happens on the field of play. But there are a lot of problems with the assist metric and how it assigns supposed value to players who accumulate them.

Firstly, there are many different definitions of an assist. It's generally accepted that an assist is the final contribution, usually a pass or cross, which leads to a teammate scoring. Many people keep records of assists, from competition organisers to journalists to fantasy games. The official Laws of the Game don't require the recording of assists; they have no impact on the final scoreline or the league table. As such, the criteria for the awarding of an assist varies. Some systems credit an assist to a player who wins a penalty or free kick for another player to convert, while others might reward an attacking player for contributing to an own goal. Some rogue systems even allow for two assists to be credited for one goal. The differences in definition between scoring systems can cause confusion.

Secondly, and more pertinently, the assist doesn't really tell us how good a player is at creating chances. A midfielder who plays a simple pass to a teammate who scores a 30-yard screamer shouldn't be credited in the same way as a midfielder who plays a defence-splitting pass which a striker is able to tap into the net. Similarly, a player might create a large number of opportunities but not reap the rewards due to the wasteful finishing of their teammates. Conventional assists data reflects just as highly on the shooter as the assister, which is unhelpful when we want to measure the creativity of players. Ideally, we want to use stats to compare apples with apples. Conventional assists stats allow us to compare apples with asparaguses.

Despite these problems, the assist still serves a purpose. The metric has been around for so long that we shouldn't just discard it. Analysts find it useful to compare modern stats to historical equivalents. Not many sports have metrics which can compare the giants of yesteryear with modern day superstars. Assists data has been retrospectively collected so that records now stretch back decades, allowing for comparisons between players like Pelé (the top assister in World Cups since such data has become available) with Cristiano Ronaldo (the top assister in the Champions League). However, there are a series of other methods and metrics which offer better insight into player creativity. Let's take a journey through the evolution of player-creativity data.

Statistical aggregators like Opta have fine-tuned the way they collect assists data. They now tag assists as either 'intentional' or 'unintentional' – was the pass meant as an assist or did it inadvertently find its way to the shooter via fortuitous circumstances? This gives a better indication of skill, while also being tagged in a way that keeps the definition consistent with historical data.

If the assist is the worm-like mammal which crawled out of the sea millions of years ago, 'chances created' is the next step in the evolution of creativity data, whereby the creature begins to grow arms and legs. 'Chances created' accounts for all passes which lead to a shot and therefore harbours no outcome bias. It doesn't matter whether or not the striker puts the ball in the net, the would-be assister still gets rewarded. That noisy, random aspect of the equation is stripped out; what is left is purely what the playmaker achieved (i.e., getting the ball into a shooting position for a teammate).

'Chances created' is a better stat than assists, but it doesn't speak to the quality of opportunities a player is forging for his teammates. Looking at the xG of the shots a player tees up gives a better view. This is the most basic form of an 'Expected Assists' model. Suppose you make a pass to a teammate and he takes a shot worth 0.40(xG), you would collect 0.40(xA). At this stage of the evolution of the player-creativity metric, the creature gains an upright posture and starts using basic tools. The primary benefit of Expected Assists is that it assesses all creative players as if they were providing for the same quality of forward. Expected Assists models are developed using a database of hundreds of thousands of shots, and xA is awarded from the probability of the average player scoring as a result of the pass. Midfielders who play for poor teams can sometimes get fewer assists, because they have worse strikers playing in front of them. If you put a midfielder in the Bournemouth team, he would get fewer assists than if he played for Barcelona. Playing for the Cherries, the forwards ahead of him would be less pacey, less skilful and have worse movement, making it harder to accumulate assists.

Suppose you feed a perfect through-ball to a striker who is one-on-one with a defender. Clearly the chance of a goal being scored from your pass is higher if Lionel Messi is the player bearing down on goal than if it is Dominic Solanke. The former is more likely to get around defenders than the latter. You have played the same pass in both situations but have a better chance of being rewarded with an assist if your teammate is Messi. In fact, you might not have even had the chance to play the pass if you were playing with Solanke, because he might not have provided the movement required for a ball to be played. (Note: Dominic Solanke is still a good footballer. He's just not as good as Lionel Messi).

The quality of striker is one noisy element which is stripped out by Expected Assists, but another is the variability in finishing. As we will see later on, the 'clinicalness' of players is inherently unpredictable. Terrible finishers can go through spells of God-like clinicalness, while the best players in the world sometimes endure periods where they couldn't hit a barn door with a banjo. Two midfielders who play in the same team and provide for the same strikers can still have very different fortunes in terms of how frequently the chances they create are being converted. The randomness in finishing ability makes assists data noisy, but Expected Assists can provide us with sound-cancelling headphones and a clearer image of who the best creators are.

Mohamed Salah amassed 13 assists and won the Premier League Playmaker of the Season award in 2021/22. However, the assists data only tells part of the story. Once we put our xA x-ray glasses on, we can see below these surface level stats and uncover his true powers of creativity. Salah's final assist of the season came in Liverpool's 2-0 win against Everton. His cross was headed home by Andy Robertson in a shot worth 0.32(xG). Salah was accredited with the assist, but he owed a lot of thanks to Robertson who converted a chance which the average shooter would miss 68 per cent of the time. Expected Assists accounts for this, rewarding Salah with 0.32(xA). Salah set up another shot 20 minutes later, this time teeing up Thiago Alcántara to try his luck from the edge of the box. This effort was saved so Salah obviously isn't rewarded with an assist, but he did manage to collect 0.07(xA) because this was the xG value he provided to Thiago. These were the only two shot-creating passes Salah managed against Everton, so in total he was awarded one assist from 0.39(xA) in the match. Adding up the value of every shot-creating pass Salah made that season leaves the Egyptian with 9.58(xA). In other words, the average forward would be expected to score 9.58 times from Salah's passes. In reality, Salah's teammates managed to score on 13 occasions from these chances – an overperformance of roughly three or four goals.

Figure 8.8 shows the top 10 assisters in the 2021/22 Premier League season, alongside their xA and xA/90 totals. The data shows Trent Alexander-Arnold to have been the most creative player in the league in terms of overall chance creation. The wing-back accumulated 12 assists but might have expected to have 13 given the 12.81(xA) he created. Interestingly, two of the top four and three of the top eight players on the list are wing-backs, which speaks to the creative responsibility of this position in modern football – particularly in a Jürgen Klopp system.

On the surface it looks like a poor season creatively for Kevin De Bruyne, who had averaged 13 assists per season since his arrival in the Premier League and who had accumulated a joint record of 20 assists just two seasons prior.

Player	Team	Minutes	Assists	xA	xA/90
Mohamed Salah	Liverpool	2757	13	9.58	0.31
Trent Alexander-Arnold	Liverpool	2856	12	12.81	0.40
Mason Mount	Chelsea	2367	10	7.05	0.27
Andrew Robertson	Liverpool	2550	10	6.24	0.22
Jarod Bowen	West Ham	3003	10	5.68	0.17
Harvey Barnes	Leicester	2108	10	4.48	0.19
Harry Kane	Tottenham	3229	9	9.24	0.26
Reece James	Chelsea	1866	9	7.09	0.34
Paul Pogba	Manchester United	1371	9	3.82	0.25
Kevin De Bruyne	Manchester City	2214	8	11.14	0.45

Figure 8.8: Assists and Expected Assists, Premier League 2021/22

However, the xA paints a different picture. Even though Alexander-Arnold clocked the largest overall Expected Assists total, De Bruyne was the player who created the most chances per minute he was on the field. The Belgian collected 0.45(xA/90) compared to Alexander-Arnold's 0.40(xA/90), meaning the Manchester City midfielder would have expected to register just under one assist every two full matches he played given the chances he was providing. The quality of De Bruyne's shot-creating passes should have seen him collect around 11 assists that season, but wasteful finishing from players like Raheem Sterling and Gabriel Jesus meant he collected three fewer than deserved. As is usually the case, though, bad luck doesn't sustain for long – the next season De Bruyne went on to create a whopping 16 assists from 16.54(xA).

On the other end of the assists luck spectrum in 2021/22 was Paul Pogba. The Manchester United midfielder's shot-creating passes accumulated to 3.82(xG), but some extraordinary finishing from his teammates meant he collected nine assists in his final season at Old Trafford.[11] In other words, even

---

[11] Remarkably, his first seven assists that season came from just 1.31(xA).

though Pogba should have been credited with seven fewer assists than De Bruyne that season, he actually collected one more than his Belgian counterpart.

No one demonstrates the regression of luck more than Mason Mount, another top 10 entry in 2021/22. The Englishman clocked 10 assists from 7.05(xA) that season, but the season prior had been the unluckiest player in the whole division, collecting just five assists from 9.57(xA). Mount created 2.52(xA) fewer in the later campaign but was rewarded with five more assists. The underperformance of 4.57 in 2020/21 was largely caused by playing behind Timo Werner, whose goalscoring exploits we'll study through an xG lens in a later chapter.

It should be becoming clear how this data can help us identify true quality. Jarrod Bowen and Harvey Barnes caused a stir in the fantasy football community during the 2021/22 season because of their sudden and unexpected bursts of assists. Many fantasy managers boarded the hype train and drafted the two Englishmen into their teams. Others were more cautious, noting that they both hadn't been creating as many good chances as it might seem and their xA overperformance would likely be unsustainable. Bowen and Barnes had collected a combined 20 assists from just 10.16(xA) and were likely to regress before too long. Indeed, the pair went on to make a combined six assists the following campaign. Whether you're a fantasy manager selecting players for your FPL team or the England manager selecting players for your World Cup squad, Expected Assists unlocks a better view of player creativity.

This analysis might prompt the reader to think, 'Hang on. Expected Assists data can clearly get it very wrong. Surely the difference between assists and xA in the cases of Paul Pogba and Kevin De Bruyne shows its uselessness as a metric? Or surely something is wrong with your model to be getting such different numbers?' But remember, the issue isn't that the xA is differing from the assists data. It's that *the assists data is differing from the xA*. The discrepancies between the two figures speaks to the luck and chance which exists within football, as per the second chapter of this book. The innate randomness of the sport is the entire point we need metrics like xG and xA. When there is a large difference between what actually happened and what would be expected to happen is the exact instance when we need these 'expected' metrics the most.

It's also worth noting irregularities between assists and xA are fairly common in one-season samples, but expanding the sample size often aligns the figures more closely. Mason Mount's assists massively underperformed xA in 2020/21 and overperformed in 2021/22, but over the two-season period he created 15 assists from 16.62(xA). To date, De Bruyne has collected 101 assists from 103.14(xA) since joining Manchester City, while Jarrod Bowen and Harvey Barnes collected 26 from 22.60(xA) over the two-year period mentioned above. Assists and xA, just like goals and xG, tend to align over long periods.

There is one more step in the evolution of chance-creation data; a creature who bears a near semblance to homo sapiens (if a little furrier and more rugged). This system, a different type of Expected Assists model, assigns every completed pass with an xA value – even ones that don't end in shots. The xA awarded to each pass depends on the finishing destination of the pass, the type of pass it was, the position of the defenders around the receiver, and a number of other factors. In this system, the majority of passes will carry incredibly low xA values. When one centre-back passes to another centre-back inside their own half, the passer might collect around 0.001(xA). The benefit of this system is it rewards creative players even if the striker doesn't get a shot away. Suppose a midfielder plays a defence-splitting pass to a forward who rounds the goalkeeper but decent control eludes him and the ball dribbles out of play. The player wouldn't collect any xA in the 'basic' version of the model because it only counts the xG created from passes. The 'advanced' model would look at the position the would-be shooter was in and credit the passer accordingly. The advanced model has close ties with another metric, Expected Threat (xT), which has been adopted by several professional clubs.

Of course, the evolution of creation data hasn't reached its final destination. There is still a huge amount of progress to be made. Every day, football analysts are coming up with new and better ways to credit actions which deserve merit, and to reward players who create goalscoring opportunities. Perhaps we're not too far away from a world where the Playmaker of the Season award is presented to the player with the largest Expected Assists total.

CHAPTER SUMMARY

- Game state is an important factor to keep in mind when analysing a football match. Teams naturally sit off and invite pressure after taking the lead. This means they tend to concede more shots, but also have the opportunity to create bigger chances at the other end.
- The way points are allocated in a league system means it's almost always worth risking one point in pursuit of three and taking risks in an attempt to win the game.
- Chance-creation data is getting more sophisticated. Expected Assists offers a better glimpse of creativity than regular assists because it assumes that all chances created are for the same quality of finisher.
- The type of analysis outlined above is influencing the way teams approach games, and altering their tactical approach to winning matches.

9

THE POSITION OF MAXIMUM OPPORTUNITY

ACCEPTING THE NEGATIVE METRIC

'Don't worry if you shoot and you miss. People will forgive you if you miss, but they won't forgive you if you have the chance to shoot and you don't'

Jimmy Murphy, former Manchester United coach

There's an area on the basketball court called 'mid-range'. This zone lies in between the 3-point line and the key (the rectangular zone around the basket). In the 2003/04 season, more than one-third of all NBA league shot attempts came from the mid-range, while fewer than one-fifth came from behind the 3-point line. The mid-range shot has suffered a dramatic fall from grace since then. By 2014, teams were taking about the same number of shots from behind the 3-point line as they were from mid-range. By 2019/20, the portion of shots taken from these positions had completely switched on its head: long-range 3-pointers now accounted for one-third of shots attempted while mid-range shots accounted for around 15 per cent. As recently as the 2015/16 season, 3-pointers accounted for more than a third of shots for only six of the 30 NBA teams. By 2020/21, 28 teams took more than a third of their shots from behind the 3-point line. The slow death of the mid-range shot is shown in Figure 9.1, which tracks the 'Rim and Three Rate' in the NBA since 2001. The Rim and Three Rate shows the percentage of all shots which are taken from either behind the 3-point line or at the rim (the rim being defined as a dunk, layup or tip-in). In other words, it shows the portion of shots which aren't taken from the mid-range.

Daryl Morey, general manager at the Houston Rockets, was one of the driving forces behind the demise of mid-range shooting. In 2017/18, the Rockets became the first team to take more than half their shots in a season from behind the 3-point line. Innovation often happens a level below the top. Consider Brighton and Brentford, lower-league football teams who revolutionised their approach in order to make it to the Premier League. The 3-point strategy was no different, emerging the level below the NBA with the Rio Grande Vipers (owned by Morey's Houston Rockets) under the coaching of Nevada Smith. 'People

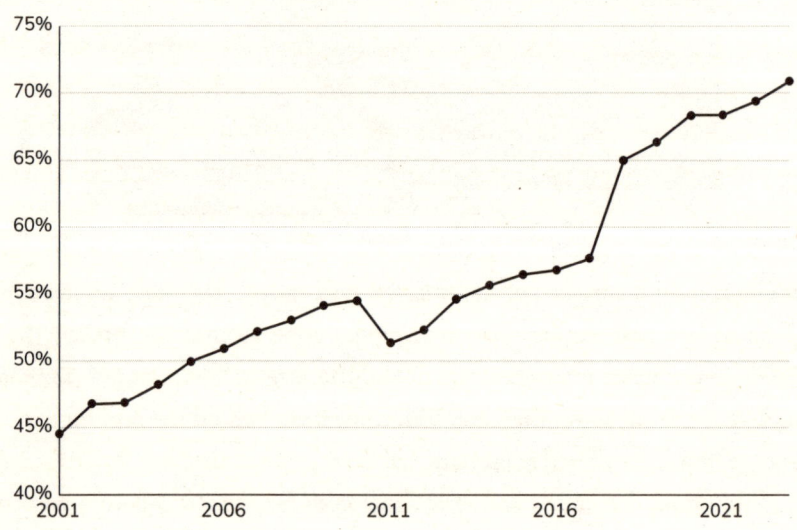

Figure 9.1: Rim and Three Rate, NBA 2001–2023

definitely thought we were nuts and ruining basketball,' Smith said. 'And now it's the most widely accepted way to play.'

It was an innovation with stunning simplicity: increasing the proportion of 3-point shot attempts in relation to 2-point attempts will increase the value of your shots in the long-run. Suppose mid-range shots worth 2 points are successful about 50 per cent of the time, we can use our trusty friend Expected Value to determine these shots are worth 1.0 expected points each. If you shoot from slightly further back, behind the 3-point line, the chance of finding the net drops to around 40 per cent but each successful attempt from this location is worth 3 points, so the expected value increases to 1.2 per shot. Over the long run, the extra 0.2 per shot will make a massive difference to the scoring proficiency of a team. You're essentially gaining 20 per cent extra value on every shot you take. Sports teams often talk about 'finding that extra 10 per cent' in terms of effort or commitment. The Rockets were able to find a material extra 20 per cent through a simple strategic alteration to their play.

The key insight behind basketball's philosophical shift was that the game had been too cautious in accepting the increased risk of missing the shot altogether. The professional poker player Caspar Berry argued, 'Whenever someone innovates in business or in life, they almost inevitably do so by accepting a negative metric that other people are unwilling to accept.' In basketball, accepting the negative metric meant being willing to miss a greater portion of shots but, ultimately, extracting more value out of the shots they did make.

Putting a value on the shots taken by either team allowed Morey and his team to look beyond the final scoreline. The Rockets were defeated by the Golden State Warriors, another leading light in the realm of basketball analytics, in Game 7 of the 2018 Western Conference Finals. Despite a 3-pointer attempt success-rate of more than one-in-three over the course of the season, the Rockets only managed seven out of 44 attempts against the Warriors. They missed 27 3-point attempts in succession, a post-season NBA record and a sequence of events with a 1-in-72,000 likelihood of occurring. Daryl Morey recognised his team had performed well and created scoring chances during the game, but luck had gone against them on this particular occasion. 'We should have won,' he stated after the match.

ShotQuality is a basketball analytics consultancy which helps teams gain a competitive advantage by providing cutting-edge insight into, well, shot quality. The company predicts shot outcomes by assigning a value to each attempt based on a number of factors – the location of the shot, the pressure applied by the defensive team, and so on and so forth. Each attempt is graded on a scale from 0 per cent to 100 per cent, representing the likelihood the shot results in a made basket. Does this sound familiar? It should do, because it's basketball's equivalent of the Expected Goals method. The big difference is the chance of success for a basketball shot is multiplied by the number of points attempted – was the effort from behind or in front of the 3-point line? This step isn't necessary in football because goals are always worth one 'point'.

Simon Gerszberg, the founder and CEO of ShotQuality, was first introduced to this xG way of thinking at Colgate University. Dave Klatsky headed up the coaching staff for the Colgate Raiders and had devised a manual system for tracking the quality of the shots that his team were taking. 'Dave would watch the footage for two hours after the game and track the percentage of each shot attempt in an Excel spreadsheet based on his intuitive knowledge of the game,' Gerszberg said. 'He made it as granular as possible based on his instincts of the players, the rhythm of each shot – and he would do it twice for each game.' In his sophomore year, Gerszberg was assigned to live grade the shots from behind the bench. He would assign each shot a percentage chance of succeeding and drop it on Klatsky's desk at the end of the game. Klatsky would go back and watch the footage, then come up with an overall Shot Quality score for the game, much in the same way a football analyst might generate an Expected Goals scoreline for a match.

'Klatsky would use that metric to evaluate the process,' Gerszberg continued. 'He'd go to the head coach and say, "Hey, I know we lost this game by 10, but we got the right shots and they just didn't fall our way." Fast-forward a few years and I was getting fed up by doing this by hand, and I wanted to watch the games! I had to find a way to automate the process of evaluating shots.' Gerszberg created a model which would collect this data for him. He quite quickly realised the value this automated system could provide to not only Colgate, but to all coaches in college basketball. He started posting metrics to Twitter – which teams were getting the best shots, which coaches were running the best plays – and his system soon caught the attention of the wider community. Before long, Gerszberg had founded ShotQuality and more than 60 NCAA teams were using his model to analyse their performances. The xG philosophy had taken hold of basketball.

COUNTERINTUITIVE STRATEGIES

In a bid to get people back in the office after Covid, my company installed a darts board and set up an intra-company competition. There were eight groups of five competitors and the top two from each group would advance to a knock-out round. Each game would be a 'best of three legs' and in each leg the competitors would work their way down from 501 to 0. My colleague Tom is an average darts player. He can get it near triple 20, the highest score on the board, more often than not. The problem average darts players face is that the '20' segment sits adjacent to both the 'one' segment and the 'five' segment, two of the lowest-scoring areas of the board. Tom wondered if there was a better way to play the game, if there was an area of the board he could target that would leave him with a larger average score per throw.

Tom can pretty accurately hit a 'slice' of three segments with every throw. The first logical step was to work out which three-segment slice added to the highest number of points. The slice which houses the triple 20 in the middle consists of the '20,' 'one' and 'five' segments, which adds to 26. That didn't seem

XGENIUS

particularly high. Tom worked out the three-segment slice with the highest combined total consists of 'seven' in the middle and '16' and '19' either side (located at the bottom left-hand side of the board). This slice adds to 42 points – 16 higher than the slice most amateur darts players aim at when they step up to the oche (the slice with 20 in the middle).

This calculation leads us in the right direction but doesn't tell the full story. When Tom aims at the '20' segment, he does manage to hit it more often than he hits either of the adjacent segments. He must bake this expectation into the equation to work out whether aiming at a different slice will actually prove more valuable. His intuition tells him he hits the segment he's aiming at around 50 per cent of the time, while he hits each adjacent segment with roughly 25 per cent of throws. With this information, he can work out the expected value of aiming at each slice:

EV = (Chance of hitting 20 × Value) + (Chance of hitting 1 × Value) + (Chance of hitting 5 × Value)

(50% × 20) + (25% × 1) + (25% × 5) = xValue of Throw

10 + 0.25 + 1.25 = 11.50

Thus, aiming for the '20' segment carries an expected points value of 11.50. This ranks fifth when compared to the 20 possible three-segment slices. The best zone to aim at did indeed turn out to be the bottom left-hand corner. The 'seven' slot carries an expected value of 12.25, the most of any slice. The worst number to aim for is 'six,' carrying just 8.75 expected points given the above workings.

The difference between the '20' slot and 'seven' slot might seem minimal (0.75 points per throw), but the impact of this difference is enhanced by the fact that doubles and triples can come into play. Once you factor in the potential to increase your score via these slots, the difference between the two slices stretches to just about one point per throw. Given it takes around 13 visits to the board for an average player to reach a reasonable checkout and each visit consists of three throws, Tom's unconventional strategy allowed

156

him to earn an additional 39 expected points per leg. Over the course of a three-leg game, that equates to an extra 117 points. Tom ended up reaching the final, playing eight matches across the group and knock-out stages in total. His strategy earned him 936 expected points over the course of the tournament – the equivalent of almost two legs. It didn't require any additional skill, nor for him to spend hours at home practising. It simply required him to aim at a different area of the board. He wasn't playing better; he was playing smarter.

Any amateur darts player would likely agree that the above strategy is optimal if you walked them through each step of the equation. Most players know they struggle to hit the '20' slice every time and lament their misfortune when a dart lands in either the 'one' or 'five' slots. Yet, interestingly, none of my other colleagues switched to this tactic. Despite recognising the dividends the strategy was paying on Tom's way to reaching the final, they branded his technique as 'anti-darts'. One commented that he was displaying 'classic Sean Dyche tendencies.' This reaction serves as a slightly abstract metaphor for sport's analytical struggles on a larger scale. The '20' slot represents the largest number of possible points, so people will gravitate towards it. They can't accept that aiming at 'seven' might be a more valid tactic than aiming at '20'. Most executives, managers and players are unwilling to disregard the obvious strategy, even when a better one presents itself. What my foxy colleague did was accept the 'negative metric.'

For NBA teams, accepting the negative metric meant sacrificing shot success rates (the percentage of shots which find the net) in return for a greater value of those which do succeed. For my colleague, accepting the negative metric meant throwing darts at a low-value number with the knowledge that that general area of the board offers greater value to the average player. Football teams have a negative metric of their own to accept: taking fewer shots in the pursuit of more valuable shots. It's essentially a reversal of the trend happening in basketball. Modern footballers are increasingly passing up long-range shooting opportunities which their predecessors would have taken without hesitation, instead attempting to work the ball into more dangerous areas. This change of tact has materially impacted the number of shots being taken – the

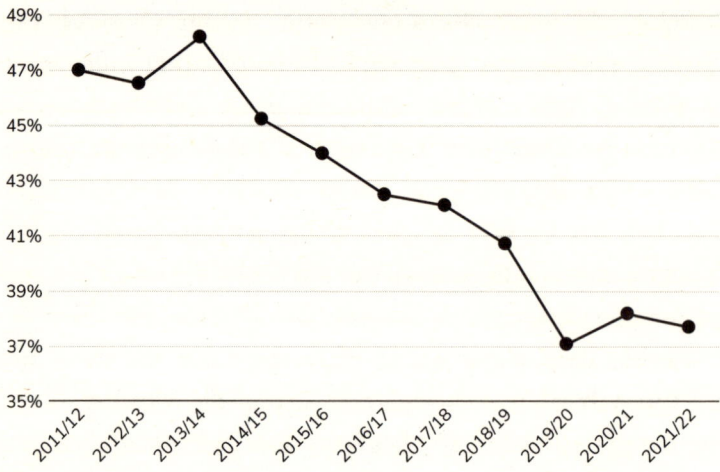

Figure 9.2: Share of Shots from Outside Box, Premier League 2011–2022

2020/21 Premier League season saw 23.9 shots per game compared to 28.4 shots per game back in 2010/11. Players are taking fewer shots, but the shots they are taking are of better quality.

Figure 9.2 shows this philosophical shift in the form of the decreasing share of open-play Premier League shots which are being taken from outside the box. In the 2013/14 season, nearly half of all attempts originated from outside the penalty area. This also happened to be the season nascent xG models began being shared among the budding analytics community. Since then, clubs have slowly but surely started accepting the negative metric. The portion of shots being taken from outside the box dropped 10 percentage points in less than a decade as teams began prioritising a low volume of big chances over a large volume of small ones. This trend is consistent across all of Europe's major leagues. The availability of Expected Goals data is having a material impact on how teams are playing the game.

THE FIVE TRUTHS OF XG

The Expected Goals method shines a light on the most important part of football: chance-creation. Teams who pay attention to xG have developed specific tactics and systems to generate as many high-quality chances as possible and concede as few as possible. This sounds obvious; surely every team tries to do this? Well, these analytical teams have implemented certain strategies to improve, enhance and boost their xG numbers. Some of these tactics are seemingly obvious, some are more innovative. They mainly stem from the five fundamental truths xG has revealed about shot-taking:

1. **The nearer to goal, the better**: This one is obvious, and we've already seen how it's affected the shooting distance of teams over the decade following the advent of xG.
2. **The more central, the better**: Shots from the central part of the penalty area are more valuable than those from tighter angles.
3. **Feet are better than heads**: All other things being equal, shots taken with feet are much more likely to result in goals than headed shots.
4. **Crosses are hard**: In general, crosses are more difficult to convert than ground passes, through-balls, and shots after dribbles.
5. **The clearer the sight of goal, the better:** Attempts with fewer defenders between the shooter and the goal are of higher value.

The first two points can be summarised by the following statement: shot location is absolutely crucial. Brentford have long since realised this. Each and every striker to have signed for the Bees in recent history has scored a multitude of goals and been sold for far more money than they were bought for. The primary reason is that Brentford teach them the fundamental principles of goalscoring: get yourself as close and as central to the goal as possible. The shot maps for Brentford strikers over the last decade or so generally show a cluster of shots in and around the six-yard box, with only a handful taking place from outside the box. Their average shooting distance from goal is among the lowest of any European team over this period.

To truly understand the importance of prioritising the creation of high-quality chances over long shots, consider Figure 9.3. This graphic shows the shooting locations of Premier League teams over the course of the 2021/22 and 2022/23 seasons combined, as well as the resultant xG of shots from those locations. Note that shots from inside the box but not in the six-yard box aren't

Team	Shots For			xG For		
	6-Yard Box	Outside Box	Total	6-Yard Box	Outside Box	Total
Manchester City	119	406	525	48.63	14.14	62.77
Liverpool	123	415	538	52.30	15.09	67.40
Arsenal	120	396	516	42.41	13.00	55.41
Brentford	115	235	350	41.18	8.20	49.38
Newcastle United	99	333	432	34.91	12.36	47.27
West Ham	97	316	413	34.37	10.23	44.60
Brighton	79	391	470	30.84	12.83	43.68
Tottenham	96	331	427	31.07	12.08	43.15
Manchester United	73	414	487	29.45	13.51	42.96
Leeds	85	344	429	30.13	10.33	40.46
Chelsea	76	369	445	26.64	13.01	39.65
Southampton	66	347	413	23.62	10.94	34.56
Everton	72	310	382	24.83	9.44	34.27
Crystal Palace	64	266	330	23.43	8.80	32.23
Aston Villa	68	325	393	20.08	10.28	30.36
Leicester	51	317	368	18.83	11.33	30.15
Wolverhampton Wanderers	40	317	357	13.03	9.62	22.66

Figure 9.3: Shooting Locations Premier League 2021/22 and 2022/23

included in the sample; for the purpose of this analysis, we're only interested in close-range or long-distance attempts. The table is ranked by the overall xG the teams created from these shooting positions.

One team immediately stands out from the rest. Brentford took 350 shots from these positions over this two-season period – the second-fewest of any team who played in the league in both campaigns. Despite this, the Bees rank fourth for xG created from these areas, behind only Manchester City, Liverpool, and Arsenal. How can this be? Well, a deeper examination of the numbers shows that Brentford were highly effective at generating efforts from within the six-yard box. The Bees regularly passed up opportunities to shoot from the edge of the box – they had 31 fewer attempts from this area than Crystal Palace, the next least prolific team from this shooting zone – and instead tried to work the ball closer to goal.

Consider Brentford's figures compared to Manchester United. The Red Devils took 137 more shots than Brentford from these shooting positions over the two-season period. In fact, United took 64 more shots from outside the area alone (414) than Brentford did from both these areas combined (350). If we break the shooting locations down, we can see that United took 414 shots from outside the area compared to Brentford's 235 – that's 179 more than their West London counterparts. But these shots only led to 13.51(xG), just 5.31(xG) better than Brentford's total of 8.20(xG) from outside the box. They may have amassed a far greater number of shots, but these efforts were all extremely low value. Meanwhile, Brentford accrued 42 more shots than United from inside the six-yard box. This may seem like a relatively insignificant number of attempts over the course of two seasons, but it meant that Brentford generated 41.18(xG) to United's 29.45(xG) from within the six-yard box, a difference of 11.73(xG).[12] Close shots matter more than shots from distance, and they matter by a much greater margin than you might think. The average shot from inside the six-yard box in the Premier League was worth 0.36(xG) over those two seasons, while the average shot from outside the penalty area was worth 0.03(xG). In other words, the generation of a chance from the six-yard box is 12 times more valuable than a pot-shot from distance.

Figure 9.4 shows the percentage breakdown of shots for each team from the same locations over that same two-season period. Brentford took the largest share of their shots from inside the six-yard box (13 per cent), while taking the fewest from outside the box (28 per cent). Interestingly, they were just as effective at the other end of the pitch. Just seven per cent of the shots Brentford faced came from within the six-yard box, joint-best with Brighton, their analytical rivals. Meanwhile the Bees forced their opponents to shoot from outside the area more than any other team in the league. Penetrating the six-yard box as regularly as Brentford were able to do, as well as being able to deny the opposition access to this area of the pitch, takes a certain level of tactical intelligence and a heavy emphasis on set-piece training. We'll study what methods the Bees have deployed to become so effective in these areas later.

The 'Justice Table', a system that ranks every team based on Expected Points at the end of the season, had Brentford as the seventh-best performing Premier League team in both 2021/22 and 2022/23, despite the club actually finishing thirteenth and ninth in the table in those respective seasons. Creating big

---

[12] Remarkably, 37 per cent of Brentford's total xG in 2021/22 came from within the opposition's six-yard box.

	Shots For		Shots Against	
Team	6-Yard Box	Outside Box	6-Yard Box	Outside Box
Aston Villa	8%	36%	8%	33%
Arsenal	10%	33%	9%	34%
Brentford	13%	28%	7%	40%
Brighton	7%	35%	7%	33%
Chelsea	7%	34%	9%	35%
Crystal Palace	8%	32%	8%	34%
Everton	8%	36%	8%	34%
Leeds	9%	36%	8%	33%
Leicester	6%	37%	9%	33%
Liverpool	9%	31%	9%	29%
Manchester City	9%	31%	10%	32%
Manchester United	7%	38%	10%	35%
Newcastle United	10%	32%	9%	36%
Southampton	7%	39%	9%	32%
Tottenham	9%	32%	7%	38%
West Ham	10%	34%	8%	36%
Wolverhampton Wanderers	5%	39%	8%	35%

Figure 9.4: Shots Taken and Conceded from Six-Yard Box or Outside Box,
Premier League 2021/22 and 2022/23

chances is one thing, finishing them is another. Unlucky bobbles, crossbar hits or exceptional goalkeeping can still undo the good work of the players and analysis departments in creating high levels of xG. Brentford scored 33 goals from the 41.18(xG) they created within the opposition six-yard box over these two seasons – eight goals fewer than expected, Meanwhile, the Bees also conceded 22 goals from 14.07(xG) from shots outside their box over this period. In other words, opposition players scored eight more screamers than expected. The data can tell you to let opponents take long shots, but it offers a cold shoulder to cry on when random variance means they happen to pick out the top corner with unusual precision.

At Brentford, the cutting-edge data and analytics is mainly hidden from the players. Deeper analysis is available to those who seek it out, but generally the guidance is boiled down into easily digestible snippets. Most notably: Don't shoot from distance! The coaches spend a lot of time talking about shooting positions and how trying your luck from a long way out doesn't make statistical sense. The players are instructed to prioritise an extra pass to a teammate in a better position rather than a long shot, and they're shown charts which outline where goals are scored from. 'It is something we talk about and coach,' manager Thomas Frank once revealed in a post-match press conference. 'Making chances bigger and put them in a better position where it is easy to hit the target. After a game when we evaluate it, we tell the players if we think they didn't make the chance bigger or reinforce the good message when they did.'

On a separate occasion, talking to Jamie Carragher on Sky Sports' *Monday Night Football*, Frank revealed another important factor which Brentford consider when assessing shots: balance. 'We're very big on playing the chance big. Don't shoot from silly areas. If you're off balance, I get mad. Why do it? Just pass it to a teammate.' The players are essentially calibrated to have subconscious xG models in the back of their heads. When they're in a game situation and a shooting opportunity opens up, their internal model will flag whether or not the effort is worth it. Tim Sparv, one of Matthew Benham's first 'data signings' for Brentford's sister club, FC Midtjylland, said, 'I would sometimes be on the edge of the box about to take a shot and I'd get an image in my head, "No, oh yeah, this is stupid shooting from here. I'm actually going to look for a pass instead." Nowadays when I watch TV and someone takes a shot from 35 yards, I always think to myself, "That is not how you score."'

An increasing number of clubs are cottoning on to this, as shown by the decreasing shot distances across Europe. Joshua Kimmich took a shot against Manchester City in the quarter-final of the 2022/23 Champions League from outside the box despite having passing options available. The camera cut to a frustrated Thomas Tuchel, who was scolding Kimmich and drawing a square with his fingers. 'He wants Kimmich to hit the target,' remarked Glenn Hoddle on commentary. In reality, Tuchel was almost certainly telling Kimmich to only shoot from inside the box. Recall Tuchel's meetings with Matthew Benham in the Smartodds offices. The German manager knows where goals are scored from. Fans often shout 'shooot' when a player is in possession of the ball from outside the box, but as the correct way of playing slowly seeps into mainstream football consciousness, will these shouts eventually be replaced with cries of 'don't shooot'?

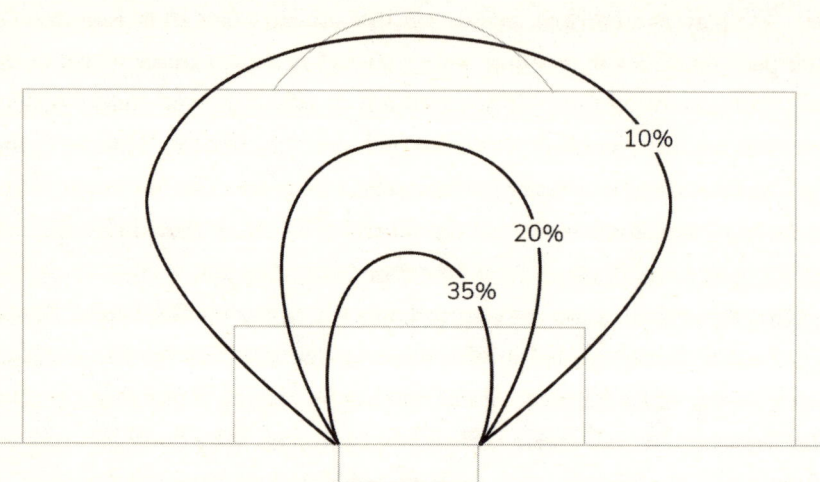

Figure 9.5: The Probability of Scoring

Creating high quality shots is one side of the equation. The other is preventing the opposition from generating big chances in and around your penalty area. Sean Dyche is the expert in this. The Englishman is often associated with 'Brexit' football and his infamous 4-4-2 formation, but he's also displayed a deep appreciation of where goals come from and how to prevent them. Dyche drills into his players the need to get back in numbers and protect the 'v' from where shots are most dangerous. This invisible 'v' effectively mimics the above visual and comes directly from the Brentford playbook of shot-generation. Figure 9.5 is a crucial visual for any player, manager, or coach to recognise. Dyche referred to this area as the 'position of maximum opportunity,' another term frequently used in the Brightonian and Brentfordian schools of xG tactics. Dyche has spoken publicly about how he demands his midfielders and wingers work back and help the defenders pack the penalty area with bodies, putting numerous obstacles between opposition shooters and the goal. 'It takes a lot of effort,' Dyche told Coaches Voice. 'But football is about effort as well as skill.' Never did a sentence so neatly encapsulate Burnley's Premier League tenure under Dyche.

Sean Dyche's Burnley went so far as to 'break' basic xG models which didn't take into account the position of defenders at the time of a shot. Such models wouldn't understand why Burnley were conceding so few goals from so many high-quality shots. The reason was they packed the space in front of goal with bodies, thus managing to block an inordinately high number of efforts. Dyche referenced the need to overload the 'second six-yard box': the zone between the edge of the six-yard box and the penalty spot. This area was identified by Brentford owner Matthew Benham in the early days of analytics as being the key scoring zone. His analysis showed that about three quarters of all goals originate from shots in this area. This zone is perhaps the most important area to control in the whole of the football pitch. Block it up. Do whatever you can to get players in the way. Crowd out that region. A consultation of Figure 9.4 shows how effectively Brentford protect their box when defending. Forty per cent of the shots they conceded in 2021/22 and 2022/23 combined were taken from outside the penalty area – the best figure in the league. Just like Dyche's Burnley, the Bees get numerous players back in defence and crowd out the primary scoring zones, meaning opposition attackers are reduced to trying their luck from distance.

These progressive clubs also don't mind conceding shots from relatively close range if the angle is tough. Consider a defender who is facing a one-on-one with an opposition attacker cutting in from a wide position. Often, an attacker can pretend to line up a shot and make the defender dive in, before skipping past him and finding himself in a much better, central position. At smart clubs, defenders are told, 'Stand your ground and let the striker take the shot. It's unlikely they'll score from wide positions, and we'd rather concede a shot from this position than let the striker inside. Shots from tough angles are fine to concede, even if it seems like they've got a clear sight of goal.' This is another way of accepting a negative metric: letting more shots take place against you from far out and wide positions in the knowledge these shots are of low value to the opposition.

Consider the recent trend of teams keeping abnormally defensive high lines when defending free kicks from around the halfway line. When sides first started doing this, it was met with bemusement by commentators. They couldn't understand what was going on and felt it was risky to leave such a large amount of space behind the defence. But the tactic has caught on and now most teams deploy high lines when defending free kicks. The thinking is to keep the ball as far away from your goal as possible. If a striker wins the first header, would you rather they do so eight yards or 18 yards from your goal? The offside rule means if you hold your line – and your nerve – you should be able to keep attackers away from the 'second six-yard box' on first contact. While it may feel safer to retreat into your own box, keeping a high line reduces the xG of any first-phase shot which might arise from the set piece. It's a tactic born from the same line of thinking as getting bodies between the shooter and your goal. Do anything you can to reduce the xG of opposition shots.

The third truth xG has revealed about the nature of shooting is that feet are better than heads. The Expected Goals method has revealed just how much we overvalue the dangerousness of headers. The reader can probably recall examples of domineering centre-forwards leaping like salmon, powerfully guiding missile-like shots off their foreheads past hapless goalkeepers and into the back of the net. These are vivid moments which easily come to mind, but it's likely the ease of recollection makes these chances seem simpler to convert than they really are. When a header is missed, commentators will say it was an easy chance and they should've at least worked the goalkeeper. Expected Goals has shown that, in reality, headers are much tougher than they look. A headed attempt will rarely exceed 0.50(xG), with most headers occupying somewhere between the 0.05 and 0.15 (xG) mark despite the fact they tend to happen closer to the goal than the average foot shot.

This relates to the fourth insight xG has given us into shooting: the difficulty of converting crosses. Crosses prompt the least valuable type of shots: volleys, first-time finishes and, on special occasions, bicycle kicks. These types of goals often make the 'goal of the season' lists precisely because they are so difficult to convert. Moreover, most crosses don't even reach any kind of attacking target at all. Only about one in every four crosses reaches a teammate. Even those pinpoint balls into the box which do find a target and, even better, find a target's feet, are still difficult to score from because they generally need to be executed first time. Roughly one in every 65 crosses produces a goal. Intelligent clubs have looked at this data and concluded that crossing is an inefficient strategy. Top teams such as Manchester City, Barcelona, Arsenal, and so on regularly ignore crossing opportunities in favour of recycling the ball and exploring other avenues towards goal. Indeed, the cross has suffered the same fate as the long shot over the last decade or so as an increasing number of teams have realised its ineffectiveness. The average Premier League side in 2010/11 attempted 16.9 open-play crosses per game, a figure which has steadily plummeted to just 12.7 in 2020/21.

However, there is one type of cross which is statistically valuable. Balls into the box from the 'penalty-area half-spaces' (as shown in Figure 9.6) are nearly twice as likely to be converted than crosses from outside the box. These balls are not so much 'crosses' as 'targeted passes' into the feet of attacking players. Cut-backs from this area into the 'second six-yard box' can generate a lot of xG. 'We've got clear principles on where we want to cross from,' Brentford manager Thomas Frank once told Sky Sports' *Monday Night Football*. 'A principle in every game we play is that the more we can hurt that half-space, the better. We want to have three or four players in the box waiting, if possible.' But Frank doesn't just want his players getting into the box, he wants them to be well-positioned when they get there in order for them to create the highest xG chances as they possibly can. 'It's very important they attack the gaps in the box when the cross is coming in. We train it, we coach it, we analyse it, show pictures. We're relentless in the coaching of it.'

Manchester City are also experts at exploiting the penalty-area half-spaces. Think about how many times you've seen balls fizzed across the goalmouth and

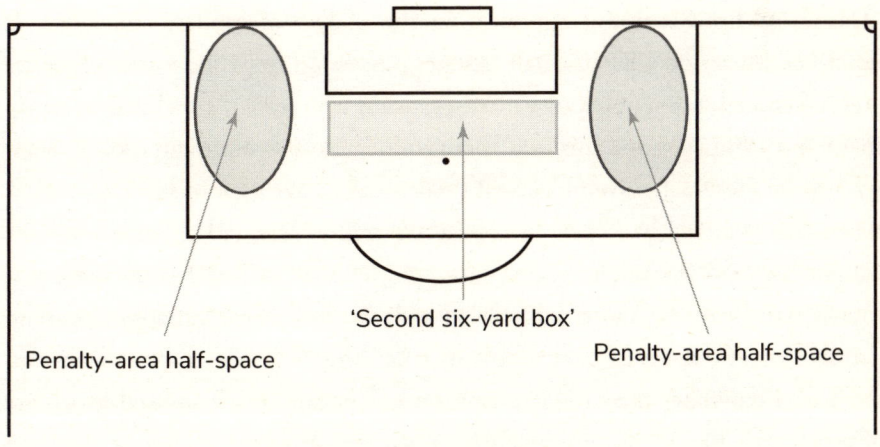

'Second six-yard box'

Penalty-area half-space

Penalty-area half-space

Figure 9.6: Critical Penalty Box Zones

players arriving at the back post to tap in from two yards out since Pep Guardiola took over. Follow Kevin De Bruyne around the pitch with your eyes for a couple of minutes and you'll likely see him attempt a dart into this area. Their rotating wingers aim to create room in these half-spaces which they can use to trouble opponents. Even when crosses from these positions don't materialise, more often than not you'll end up dragging a centre-back out of position to cover the half-space, which should theoretically leave you with a numerical advantage in the middle of the box once the ball gets recycled.

At the other end of the pitch, Brentford are happy to allow the opposition to make crosses into their own box. Since promotion to the Premier League, the Bees' most commonly used formation has been a 3-5-2 which serves as a 5-3-2 in defence. This allows them to congest the central areas of the pitch with three centre-backs and three midfielders screening in front. They concede space out on the wing in front of their wing-backs, but they're happy for the opposition to have possession in these areas. Balls into the box from these flanks are easily dealt with by their three commanding centre-backs, and the midfielders work hard to clog up the 'second six-yard box' when they're under the pump. All these strategies are geared around one thing: preventing the opposition from accumulating xG.

As well as minimising the amount of space your opposition can work in, low blocks can also help with chance-creation in transitional moments. 'Transition' has become a key term in the dialect of modern managers. It describes the period when the ball changes possession from one team to another and is a crucial moment because often the team who loses the ball will be set up in their attacking shape, making them vulnerable to counterattacks. A team who sit in a low defensive block will theoretically have a lot of space to exploit when they win the ball back as opposition midfielders and defenders will be caught high up the pitch. The chances they're able to create from these fast breaks will likely be more valuable as well because these situations tend to produce more one-on-ones or moments when there's less defensive pressure on the ball. Remember the earlier section on game state which showed teams are able to create higher xG opportunities when they're winning because they can pick opponents off on the counter. A team who utilise a low block are accepting a volume disparity: they will likely concede a lot of long shots, but in turn will create a few big chances at the other end. In a sense, operating a low block is accepting two negative metrics: concede more chances but of low danger, and create fewer chances but of high danger. Hopefully the xG you create from your high-quality chances will outweigh the xG your opponents create from their low-quality ones.

Roberto De Zerbi wanted the best of both worlds. When he arrived at Brighton, he realised that letting Premier League opponents operate in your half of the pitch is dangerous. Even if you clog the area up with defenders and deny them space, elite clubs have the quality of talent to hurt you regardless. De Zerbi wanted to find a way to create big chances in transition while also controlling the game.

Roberto De Zerbi's tactical set-up bears similar characteristics to that of William the Conqueror.

His solution bears a striking resemblance to the strategy used by William the Conqueror at the Battle of Hastings in 1066. William's Normans were faced with a shield wall of tightly packed English soldiers on a ridge at the top of a hill. Harold Godwinson, leader of the English, was effectively deploying a low block. He hoped the Normans would make repeated assaults on the unbreakable shield wall, wearing them down before counterattacking to finish them off. Harold would have got along with Sean Dyche. At first, the Norman attacks proved fruitless against the might of the shield wall. Then a rumour broke out among the French soldiers that William had been killed. General panic broke out in their ranks which sparked a retreat. Encouraged by the fleeing of their opponents, a handful of Englishman broke the shield wall and chased the French down the hill. William wasn't in fact dead and watched as his Normans picked off these overeager Brits. This lit a lightbulb in his head. The French made a strategic change: instead of launching a series of attacks trying to break the shield wall, they would instead launch attacks and then feign panic and pretend to flee, luring Harold's soldiers out of the shield wall and down the hill where they could be picked off. This tactic led the Normans to victory and ended over 600 years of Anglo-Saxon rule.

Roberto De Zerbi's tactical set-up bears similar characteristics to that of William the Conqueror. Brighton bait opposition teams into pressing them high up the pitch. Brighton's centre-backs often stand still with their studs on the ball on the edge of their box, trying to provoke the other team's strikers into pressing them. In turn, the opposition move up and open space in midfield or in behind, creating a situation where 'artificial transitions' are available.

But it's not quite that simple. Although Brighton will have lured the opposition up the pitch, the space around their centre-backs on the ball will now be tighter. One solution is to hit the ball long and behind the now advanced opponents for strikers to chase, but these can often be easily dealt with by opposition defenders and sweeper keepers. Brighton seek higher-value transitional moments. They want to play the ball into the midfield and use a series of short passes to cut through their opponents like a knife through warm butter. Brighton's strikers will often drop deep to join in the build-up play, either outnumbering the opposition in midfield or pulling their markers with them, and have their wingers make runs into the space vacated by opposition centre-backs.

Brighton centre-back Lewis Dunk opened up on how his team play out from the back. 'We know where the pressure is coming from, so we know where the options are,' said Dunk. 'The manager is relentless every day in training to work on how we handle different pressures. We're calm when we're doing it and we know we've got two or three options. We work on patterns every single day in training and the manager wants perfection. We've got a set pattern for every time we get the ball. When I get pressed in one situation, I know what the pattern should be for that specific situation.' The risk of such an approach is high, but so too is the reward. Being able to consistently execute these 'patterns' takes an immense amount of training-ground practice.

Brighton have essentially broken the fluid game of football down into a series of predetermined moves. They've devised routines in a similar way to a basketball or American football team, or how a set-piece coach might plan sequences for a corner kick. Playing this style also requires a concerted recruitment drive to acquire players with the technical ability to execute such manoeuvres. No doubt Brighton players are also told not to shoot from distance and that exploiting the penalty box half-space as much as possible is the most statistically valuable way to play the game of football. Analytical clubs have turned the sport into a game of chess, whereby they move pieces about, trying to get into the highest-value areas and bait opponents into moving out of position.

CHAPTER SUMMARY

- Multiple teams across several sports are learning to accept a negative metric, which means they have found a better, yet seemingly counterintuitive, strategy or approach.
- Football teams are taking fewer and fewer shots, but the shots they are taking are of higher quality.
- The easily detectable downward trend in shots from outside the box can be attributed to the increased use of xG analysis by clubs.
- Innovative clubs are paying close attention to chance creation, and rehearsing what sort of opportunities they want to carve out. For instance, they are looking at what types of crosses generate the best goalscoring situations with the number of crosses in general also declining sharply over the past decade.
- Expected Goals is influencing tactics and helping forward-thinking clubs master the science of winning matches.

10

THE CORNER FLAG

THE ART AND SCIENCE OF DEAD-BALL SITUATIONS

'Set pieces are a game within a game'

Gianni Vio, former Tottenham Hotspur set-piece coach

Jürgen Klopp was not happy. His Liverpool team had just been defeated 3-1 by Brentford at the Gtech Community Stadium and Klopp knew the reason why. 'The set pieces allowed Brentford to create chaos. That's what they do. They do it really well. I respect it a lot, it's really well organised.' The German shouldn't have been particularly surprised by either the result of the match or the manner in which Brentford earned their victory. Three games earlier the Bees defeated Manchester City at the Etihad, the only team to do so that entire season, scoring their first goal from a free kick. The game after that they drew 2-2 with Tottenham, their second goal coming from a corner. They then defeated West Ham 2-0 at London Stadium, both goals coming from throw-ins. Little old Brentford, despite possessing the smallest playing budget in the division, had taken 10 points from a possible 12 against Manchester City, Tottenham, West Ham, and Liverpool. And they'd scored set piece goals in each of those games.

Liverpool have long since understood the importance of set pieces. They took the 2018/19 title race down to the very last day by virtue of scoring 22 set-piece goals to Manchester City's 12. Without those 10 extra goals, Liverpool's title charge would have faded long before the final day. The stark contrast between the set-piece proficiency of the sides made Manchester City sit up and take notice. They recognised the need to improve in this area and, through a recommendation from assistant manager Mikel Arteta, hired Nicolas Jover as their set-piece specialist in the summer of 2019. Jover ended up joining Arteta at Arsenal two years later and helped the Gunners record their highest return for goals scored from set pieces in a Premier League season (15) since records began in 2006/07. Just as Liverpool stayed in the 2018/19 title race due to set-piece proficiency, so too did these goals help Arsenal to their joint best league finish in nearly two decades in 2022/23. If set-piece goals are important at the top end of the table, they are even more so at the bottom end where goals are rarer and thus more valuable commodities. The most obvious decision a relegation-threatened team can make is to train set pieces like the world depends on it.

In every football match, there are two concurrent battles running alongside one another: the 'open-play battle' and the 'set-piece battle'. The former of these gets most of the attention. After all, around two-thirds of goals are scored from open play. When we think of a game of football, we think of two teams going head-to-head out on the open field, passing the ball, making tackles, dribbling past one another. But there's another battle going on, one which takes place from corners, free kicks, and throw-ins. The 'set-piece battle' is often forgotten about. It doesn't tend to get coverage in the post-game analysis unless one team makes a disastrous error from a set piece, but it's a massive edge which progressive clubs exploit. Only about 1.9 per cent of set pieces lead to goals, a stat which some coaches believe proves that practising them is a waste of time. However, the average open-play possession only leads to a goal 1.1 per cent of the time, meaning set pieces are almost twice as valuable as having regular possession of the ball. Roughly 60 set-piece situations present themselves to each team in every game, which means 60 opportunities to cause chaos and create a chance. Moreover, set pieces are the easiest part of the game to practise and rehearse. They offer an environment which can be controlled. Routines can be devised to create space in the opposition box for your players and increase your xG figures.

Figure 10.1 shows the number of corners each Premier League team took across both 2021/22 and 2022/23, ranked by how frequently they were able to create a shot from those corners. Brentford came top of the table, managing to produce an impressive 168 efforts at goal despite earning the second-fewest number of corners of any team (322). Incredibly, more than half of their corners resulted in a shot over this two-season sample. Newcastle and Liverpool also made the top three, both of them being smart clubs who realised the opportunity

Team	Corners	Shots from Corners	Corners Resulting in Shot
Brentford	322	168	52.2%
Newcastle United	430	215	50.0%
Liverpool	517	250	48.4%
Leeds	369	178	48.2%
Southampton	390	172	44.1%
Arsenal	431	190	44.1%
Brighton	433	186	43.0%
West Ham	406	172	42.4%
Manchester City	554	234	42.2%
Everton	336	137	40.8%
Crystal Palace	361	144	39.9%
Tottenham	396	156	39.4%
Manchester United	393	154	39.2%
Wolverhampton Wanderers	353	138	39.1%
Chelsea	450	170	37.8%
Leicester	320	119	37.2%
Aston Villa	359	131	36.5%

Figure 10.1: Premier League 2021/22 and 2022/23

afforded by set pieces. It's striking how poor some of the big teams were at generating shooting opportunities from corners: Chelsea, Manchester United and Tottenham all ranked in the bottom six. With the resources these clubs have at their disposal, they should certainly be able to afford a specialist coach whose sole purpose it is to maximise their chance-creation from dead-ball scenarios. Perhaps there is an arrogance at play here. Maybe the head coaches don't think they need to waste time on set pieces, time which could otherwise be spent on tweaking advanced tactical systems. Or perhaps the superstar players shun the idea of standing around in the wet and the cold as a set-piece coach instructs them how best to attack a cross swung in from a corner. Either way, these teams are missing a trick. Their neglect of set pieces may be a contributing factor as to why they've fallen behind the likes of Manchester City, Liverpool and Arsenal in recent years – all of whom have a better appreciation of the need to execute set pieces well.

Since Brentford discovered the massively undervalued nature of set pieces, they've hired a series of specialist coaches to devise routines and make them the best in the business in this area. The aim is to win the set-piece battle in every single game, accumulating more xG from these situations than their opponents. Brentford's success in this area has been so profound that four of their five set-piece coaches have since joined other Premier League clubs, including Nicolas Jover (Arsenal) and Gianni Vio (Tottenham). Brentford also brought in former NFL and NBA coaches to consult on play design. The predetermined nature of set pieces means they bear a resemblance to an American football play or basketball out-of-bounds routine. Brentford even have a WhatsApp group containing the owner, members of the management staff and coaches in which they send ideas for innovative routines and clips of any unique set-piece sequences deployed across European football.

Taking advantage of the corners you win is one thing, but Brentford have actively started devising strategies to earn themselves more of these situations in the first place. The manager tells the left centre-back to launch long balls down the left channel in the knowledge that the opposition right-back will likely be forced into conceding a corner or throw-in on which they can capitalise. These aren't just hopeful hoofs forward: they are targeted long balls prompted by a predetermined run from their left-wing-back and supported by a rehearsed counter-press. The primary purpose of this move is to generate dead-ball situations for themselves, whether that be a corner or a long-throw opportunity.

The fact that Brentford have been possibly the best dead-ball team in the world is no accident. The ownership have long since realised the difference set pieces can make on the success of a team. They even go so far as to incentivise the players and coaches financially via a set-piece clause in their contracts. Every time the team score from a corner or free kick, they earn a special bonus. This motivates the management and playing staff to practise and execute these situations to the best of their ability. They were the first club to do this, and are still one of a select few to incentivise set-piece output. The management staff even keep a tally of the set pieces Brentford have scored from on a white board in the canteen where the players eat.

Brentford estimate their set-piece expertise grants them an additional $0.50(xG)$ differential per game over their opposition. In other words, they generate something like $0.25(xG)$ more and concede $0.25(xG)$ less than they otherwise would as a result of their extra focus on this area of the game. That means they gain a positive swing of roughly $19.00(xG)$ per season. For context, Harry Kane averaged $19.34(xG)$ per season during his spell at Tottenham. Brentford have essentially added a Harry Kane-like presence to their team through simply spending more time thinking about and practising set pieces. They have all but signed a world-class striker at the price of a set-piece coach. It's unlikely Bernardo Cueva, Brentford's latest set-piece mastermind, was on the £10.4 million per year Kane was paid.

An obvious way in which Brentford gain an advantage from set pieces is by closely studying their opposition's weaknesses from such situations. The first step to a successful set-piece strategy is to understand the Achilles heel of your foe. When Brentford plan for battle, they spend hours watching footage of their opposition defending set pieces. They will note any big xG chances they concede and, crucially, how those chances came about. Perhaps they struggle to defend in certain areas or have a vulnerability against certain runs. Before the match against Liverpool outlined at the beginning of this chapter, Brentford spotted the Reds' weakness at defending front-post deliveries. The first goal of the game came from such a routine, and they had another goal ruled out for a marginal offside after a front-post corner was flicked on. Each set piece Brentford deploy in any given game is designed to exploit the opponent's flaws.

The Bees also don't see set pieces as isolated events, but as a series of routines they can plan as a whole for maximum effect. After a number of corners fired into the near post, Brentford caught Liverpool out by delivering one to the back post. The unexpected nature of the delivery was such that Yoane Wissa had time to chest the ball down and smash it into the net from 10 yards out (although this goal was also ruled out by VAR because of a marginal offside). Brentford have multiple routines which they can play from the same initial set-up, meaning the opposition never knows which ones they'll be defending against in any given situation. Deception is key to the Bees' set-piece success. They often deploy a swarming technique in the box which allows them to conceal each routine until the last minute and confuses the opposition. To do this, they put a large number of outfield players in a small space in the opposition box to make it impossible for the defending team to read what their individual roles or runs are likely to be. Just before the kick or throw-in is taken, the players will fan out and create space for one another with various predetermined runners and blockers.

Each set piece Brentford deploy in any given game is designed to exploit the opponent's flaws.

Timing is another crucial aspect of Brentford's dead-ball skill. First of all, they take their time in taking any corner, free kick or throw-in they might win. They don't rush, take time to get players into position and build the pressure on the defending team. When it comes to the actual execution, they make sure they're the protagonists of the situation. They ensure the ball and their players end up in the right place at the right time, and that their opponents are always a step behind. They'll often manipulate the timing of their routines by playing corners short, meaning their players have time to get free from their men in the box before the ball is eventually delivered. The delay makes it much harder to defend against.

Brentford are always innovating from set-piece scenarios and looking for new ways to get ahead. One particular innovation borne out of data analysis which immediately catches the eye is that Brentford leave no players back from attacking set plays. Every Brentford player enters the opposition third and the deepest defender can be found at the edge of the opposition box. Brentford's tactical set-up at these dead-ball situations goes against decades of set-piece theory. It looks like a crazy approach when you see it in action. It seems as though there's an immense amount of space for the opposition to break into but, having crunched the numbers, Brentford found that only 0.2 per cent of attacking corners are conceded from – that's one in every 500. To date, Brentford have only conceded one goal as a result of pushing all their players up at corners, and this was against Bournemouth in a Championship play-off semi-final when they'd only just adopted the approach and hadn't finessed the transition back into defence if the opposition did break quickly. When questioned on the tactic, manager Thomas Frank said, 'Some people call it risky, but if you're not taking risks then you're also taking risks.' In a way, this quote sums up Brentford's whole philosophy of invention and innovation. They're not afraid to try new things, taking calculated risks to eke out whatever edges they can, going against the grain and creating a number of small margins. A large number of small margins going in your favour will make a big difference.

The benefits of pushing everyone up at corners are twofold. Firstly, packing the opposition box is a surefire way of creating more xG. More attackers in the danger area means more chance of the ball falling to one of your players to get a shot off. Secondly, pushing everyone up means the ball is more likely to stay in the opposition third of the pitch. Even when the opposition do manage to get it clear, the lightning-quick fullback will mop up and pass back to the goalkeeper. Brentford will leave their main aerial threats up the pitch and these players will pull out to either wing for Brentford's goalkeeper to pick out with his exceptional long-passing ability. The aerially dominant players will most likely beat the opposition fullbacks in the air and head the ball back into the box, meaning the second ball once again comes up for grabs in the opposition penalty area. The second phase creates more dangerous situations for Brentford, particularly as the defending team will be less organised in this phase than they initially were from the corner.

Gary Neville said on commentary during the win over Liverpool, 'Brentford look so dangerous from corners. It looks like they've got 15 players in there.' What Brentford's strategy ultimately boils down to is trying to get as many men as possible in scoring positions, while getting as many men as possible in front of your own goal when defending. The tactic essentially allows you to play with an increased number of attackers and defenders. What happens in the middle of the park is relatively inconsequential: all that matters is the way the odds fall when you're in either box, particularly the 'second six-yard box'. Football is a game of chess and Brentford are masters at getting their pieces in the right positions. Their ability to get bodies in either penalty area allows them to create more dangerous and concede less dangerous scoring opportunities. These insights were born out of xG. In 2022/23, Arsenal began deploying the Brentford tactic of throwing every player forward at attacking dead-ball situations on their way to their most successful set-piece season since records began.

Another easy benefit Brentford have found from corners is bringing over a second 'taker' to stand over the ball. As well as masking what routine they're about to deploy, this forces the opposition to decide whether to send a man out to mark the spare man. If they don't, then a short corner would allow the Brentford corner takers to run into the box unopposed. But if the opposition do send a man out, which they usually do, it draws a defender out of the penalty area. Sending another Brentford player out of the box as a second passing option means yet another defender will likely be lured out of the box. It's a game of numerical advantages and trying to weigh the odds as much in your favour as possible. Drawing opposition defenders out of the penalty area at set pieces means your players are more likely to get on the end of the ball and there will be fewer bodies to shoot through if they do.

In terms of actual set-piece routines, Brentford have formed a repertoire of different moves and sequences designed to bamboozle their opponents. They use clever runs, blockers, and decoys in order to create space in the area for their players. Most of the strategies are aimed at creating chances from the second ball. The theory goes that the opposition is often well set up to deal with the first ball into the box, but, for example, playing a ball to the back post to be headed back across goal creates a great deal more chaos and means their attacking players can get free more easily. The same goes for front-post flicks. And it's not just corners which Brentford have put under the microscope; they also have a range of free-kick routines in the locker which aim to catch the opposition out.

Additionally, they've brought the long throw back into vogue. Synonymous with Tony Pulis, Stoke City and Rory Delap in the seasons after the club's promotion to the Premier League in 2008, the long throw has served as a great weapon for Brentford in creating 'organised chaos' in the opposition penalty area.

Domestic football is the typical breeding ground for innovation. The amount of wealth being pumped into the sport means finding those small edges has become increasingly valuable. International football, on the other hand, pales in comparison to the invention, technical quality, and tactical astuteness of the domestic game. Champions League and Premier League matches could be considered the pinnacle of the sport, with World Cup and European Championship matches being some way off that level. International teams simply don't have the same amount of time to practise various strategic approaches. This time scarcity has led to one innovation though: national teams are doubling down on their set-piece approaches. Nations are seeing dead-ball scenarios as an easy way to make their teams more efficient and generate more high-quality chances: 43 per cent of the goals scored at the 2018 World Cup came from set pieces, the largest share since 1966. England manager Gareth Southgate studied NFL and NBA play-design in advance of the tournament, and the Three Lions went on to score nine set-play goals in Russia – more than any World Cup side for over 50 years. Three years later, at the postponed Euro 2020, shock semi-finalists Denmark and tournament winners Italy both found success with the help of set-piece coaches who had previously been employed by Brentford.

SPECIALIST COACHES

There are three key stages to the data disruption of any sport. The first is finding better players. We have seen how xG can be used to make better recruitment decisions and identify hidden gems. The second is playing the game better. Hopefully these last few sections have demonstrated how analytics has guided playing style and brought set pieces to the fore. The third is training players better. This last stage is often underappreciated when talking about the data revolution of football. Arsène Wenger was the first to make strides in this field, introducing new training methods which focused on enhancing his team's speed, agility, and endurance. He recognised the importance of recovery in preventing injuries and introduced new techniques like cryotherapy and massage therapy. He also forced his Arsenal team to drink less beer and eat less pizza. He hired a full-time nutritionist to work with the players and design personalised meal plans for each individual based on their specific needs and requirements. Wenger's revolutionary approach allowed Arsenal to complete the entirety of the 2003/04 Premier League season undefeated. Soon enough, the Frenchman's training and match preparation methods became standard practice throughout the Premier League.

Analytics has struggled to gain traction when it comes to player training, predominantly because of the type of people who are empowered in these roles. An increasing number of recruitment teams are hiring from non-football backgrounds – it's common to see ex-city traders or PhD mathematicians having a say in player acquisition. But coaching is still very much in the dominion of ex-professional footballers who are less likely to have an academic background and therefore less likely to adopt an analytical approach. They are also less likely to innovate, having played the game in the same way throughout their careers.

Brentford sought a different path. They hired a series of experts from outside the world of football who helped them challenge conventional wisdom. Just like the Oakland A's discovered the undervalued nature of deception in the market for pitchers, Brentford realised the hidden value of deception in the way footballers kick the ball. Specialist 'kicking coaches', most notably Bartek Sylwestrzak, have worked with Brentford's squad in the same way swing coaches work with golf players, with a particular emphasis being put on how to kick the ball with topspin and minimal back-lift. Emiliano Marcondes was signed from FC Nordsjælland because of the way he kicked the ball with incredible dip and movement. Marcondes scored in the Championship play-off final, the game where Brentford won promotion to the Premier League, with a trademark strike that saw the ball dip into the bottom corner.

Brentford also employed the services of high-performance sleep consultant Anna West. 'Our philosophy is to focus on doing the basics incredibly well in order to maximise results,' explained Brentford's Head of Performance, Chris Haslam. 'Besides training, we truly believe quality sleep is the biggest fundamental tool a player can use to reach peak performance on a daily basis.' Good sleep is incredibly important to both a player's physical and mental wellbeing. West taught the Brentford players that we sleep in 90-minute cycles. 'During the first part, deep sleep is dominating, and this is what recovers you from a physical perspective,' she said. 'This means the percentage risk of injury increases significantly if a player is not sleeping well. In the last part of sleep, the upper layers are dominating, and this is very important from a cognitive, emotional and psychological perspective.' In essence, sleep has a big impact on learning ability. A bad night's sleep might make a player have to learn something 10 times the next day instead of five times in order for it to stick.

Just as Arsène Wenger's nutritionist worked one-on-one with the Arsenal players to develop bespoke dietary plans, West curated specialised sleep plans for the Brentford players. She carried out an initial screening of the players by making them fill in a questionnaire – do they have children, how often do they stay in hotels, when do they eat, how much do they sweat – and then divided the squad into a set of traffic-light profiles. Reds are most prone to problems falling asleep, yellows were slightly better, and greens were likely the best sleepers. West's work with Brentford has had a big impact on the squad's on-pitch performance, injury record, and ability to retain tactical information.

Brentford also employed an even more niche specialist coach. Thomas Gronnemark boasts an eclectic CV: sprinter, Danish bobsleigh team member, and professional throw-in coach. It was the latter of these talents Brentford made use of when in the Championship. Gronnemark puts throw-ins into three categories: long, fast, and clever. Brentford were mainly interested in long throws, which they saw as a valuable weapon for chance-creation. Gronnemark coached Brentford centre-back Mads Bech Sorensen to throw the ball over 40 metres while maintaining a flat trajectory which lent itself to dangerous flick-ons. He also worked with Brentford's sister team, FC Midtjylland, who scored a remarkable 35 goals from long throws between 2015/16 and 2018/19.

Even after Gronnemark's departure in 2020, Brentford continued to practise throw-ins in training at least once a week. They've used Gronnemark's methods to teach diminutive central midfielder Mathias Jensen to launch the ball deep into the opposition box and set-piece coach Bernardo Cueva devises corner-like routines to cause chaos and create chances from long throw-ins. The fruits of Brentford's labours are clear to see. During the 2021/22 campaign, the Bees averaged just under one shot per 90 minutes from throw-ins. Every other team in the Premier League put together averaged just over one shot per 90. Brentford created 0.10(xG) per 90 from throw-ins while all other 19 teams combined created 0.12(xG) per 90. In other words, Brentford would be expected to score 3.80 goals per season from throw-ins, while the average team would be expected to score 0.24. If Premier League goals are estimated to be worth around £2 million per unit, the impact of Brentford's long-throw focus was around £7m. The wages of Premier League backroom staff members are undisclosed, but it's safe to assume the appointment of a throw-in coach represents a reasonable return on investment.

After leaving the employment of Matthew Benham, Gronnemark joined perhaps the biggest English team who have put innovation at the centre of their club strategy: Liverpool. The Reds were more interested in working on the fast and clever throw-in categories, caring less about long throws into the box and more about retaining possession from the 50 or so other throw-ins which occur each game. Gronnemark did help Andy Robertson increase his throwing distance from 19 metres to 27 metres, but his biggest achievement was to increase the squad's 'throw-in intelligence'. 'It's not about memorising a playbook,' Gronnemark stated. 'It's about helping the players think and work through the solutions themselves.' Gronnemark's spell at the club helped Liverpool win the Premier League and the Champions League.

Liverpool, themselves, have been known to draw upon expert wisdom from beyond the realms of traditional football coaching. Before their title-winning season, Jürgen Klopp invited renowned high-wave surfer Sebastian Steudtner to the club's pre-season training camp. Steudtner worked with the players on holding their breath underwater with the aim of retaining focus and calm under pressure. The best at the task were able to fully submerge themselves for little more than a minute when he arrived, and some could only manage around 20 seconds. By the end of the session, Mohamed Salah and Dejan Lovren were approaching the four-minute mark. The results of Steudtner's visit aren't easily trackable, but Liverpool did manage to win 14 games by a one-goal margin that season on their way to a first Premier League title.

Sleep coaches, kicking coaches, throw-in coaches. These job titles attract sneers from traditionalists. 'I want to be the first kick-off coach,' Andy Gray sniped when discussing Thomas Gronnemark's work. (Side note: a handful of teams have actually targeted kick-off as an area of the game worth practising. Bournemouth scored the second-fastest Premier League goal ever against Arsenal after just nine seconds in March 2023, having scored another rehearsed kick-off goal 15 months earlier against Fulham). Despite the derision from 'old school' football types, the importance of exploiting edges cannot be overstated. Progressive clubs are constantly innovating and seeking out marginal gains which can improve the performance of their team.

CHAPTER SUMMARY

- About one-in-three goals are a result of a set-piece situation. Set pieces are also the only moments in a game that a team can have full control over and practise routines for.
- Set pieces allow teams to pack an opposition box, automatically creating more xG.
- Historically, teams and coaches haven't practised set pieces as much as they should have done given their importance.
- While this is changing, dead-ball situations still offer an important edge for clubs looking to innovate or compete on tighter budgets.
- Specialist coaches are another such edge that teams can gain in their mission to win more matches, whether that be in helping train corners, throw-ins, sleep, and so on.

11

THE BACK OF THE NET

THE RANDOMNESS OF CLINICALNESS

'I hear people talk about Andy Cole and say: "But he needs two or three chances before he scores". Well, if he's getting that many chances, I want him in my team'

Les Ferdinand, football coach and former professional player

Mason Mount fired in a shot from 25 yards which arrowed towards the bottom corner of the goal. A scrambling Łukasz Fabiański managed to get a hand to the ball but couldn't divert it around the post. The ball instead fell to the only player who had followed the shot in, Timo Werner. Inside the six-yard box and with the goal at his mercy, this was a position Werner was all too familiar with that season. And a familiar outcome occurred, as his weak left-footed strike across the goal dribbled despairingly wide.

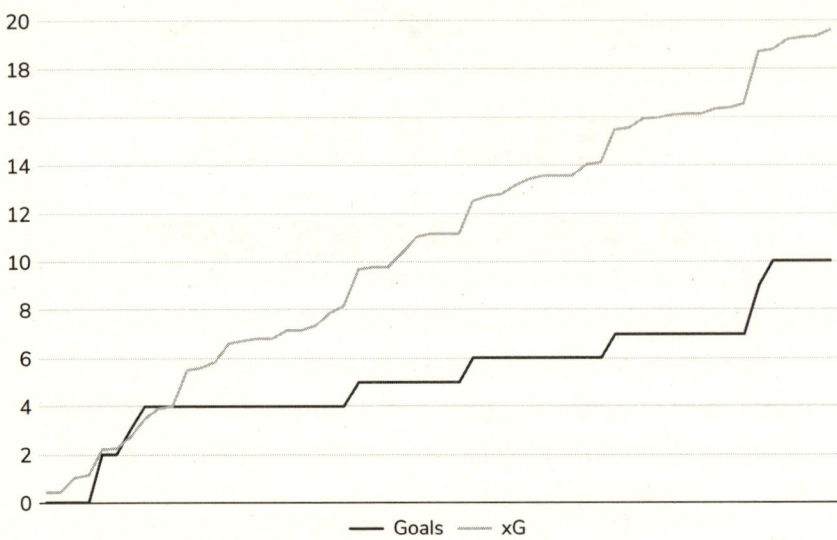

Figure 11.1: Timo Werner – Goals vs xG, Chelsea Spell

This attempt, worth 0.63(xG), was the last of the 18 'big chances' Werner missed during the 2020/21 League campaign.[13] He missed 78.3 per cent of the 'big chances' presented to him, a figure bettered (or should that be worsened) by only five Premier League players on record. Werner scored six goals from 13.54(xG) in the Premier League that season, meaning he found the back of the net on about seven or eight fewer occasions than the average player would have been expected to given the chances he was presented with. Werner's xG deficit has gone down in the annals of football history as one of the worst underperformances since the advent of Expected Goals. Figure 11.1 depicts the cumulative goals and xG he registered during his time at Chelsea from 2020 to 2022. Overall, he scored half as many goals in his spell at Stamford Bridge as he'd be expected to. Werner's wastefulness became a sensation on social media. Fans were watching him squander chance after chance, and the xG data was backing up the fact that Timo Werner was a terrible finisher.

---

[13] Stats Perform define a 'big chance' as a situation where a player should reasonably be expected to score, usually in a one-on-one scenario or from very close range when the ball has a clear path to goal and there is low-to-moderate pressure on the shooter.

Except, was he? In the six seasons before he joined Chelsea, Werner scored 87 goals from 83.86(xG). The season prior to his transfer to England he scored 28 goals from 22.23(xG) for RB Leipzig in the Bundesliga, netting on roughly six more occasions than you'd expect him to. Figure 11.2 shows Werner's rolling goals and xG figures in the League from the start of the 2014/15 campaign up to the end of his Chelsea spell. The xG output remained fairly consistent throughout this period, but the number of goals he scored was erratic. He went through hot streaks and cold streaks, none more so than the purple patch he experienced at Leipzig in 2019/20 which preceded his barren run at Chelsea the following campaign. One season everything Werner touched turned to goals, the next season everything he touched turned to wasted xG. How can the same player produce such contrasting xG numbers?

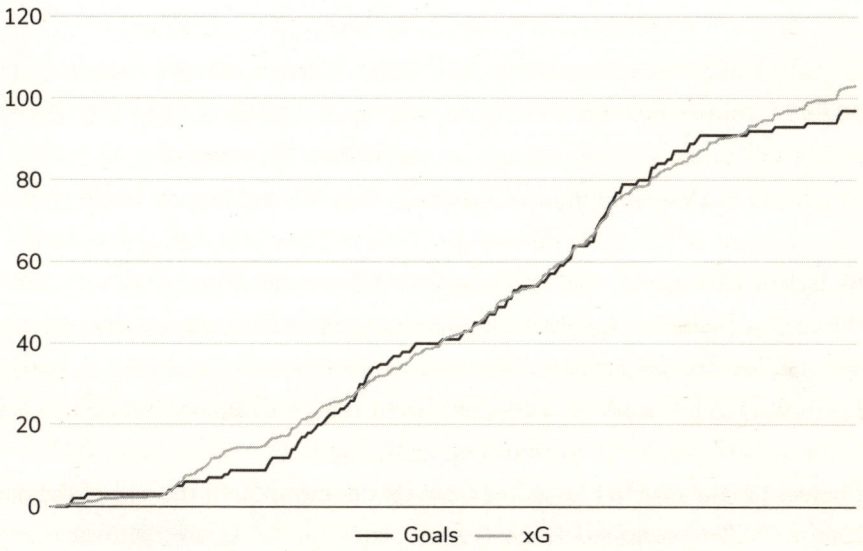

Figure 11.2: Timo Werner – Goals vs xG in League, 2014–2022

Another player was also struggling to find the back of the net at the same time as Werner during the first half of the 2020/21 season. Kevin De Bruyne had developed a Werner-like habit for squandering opportunities, as shown in Figure 11.3. The Belgian scored just three times from 8.58(xG) during the first 22 matches of 2020/21. His xG underperformance over what amounted to

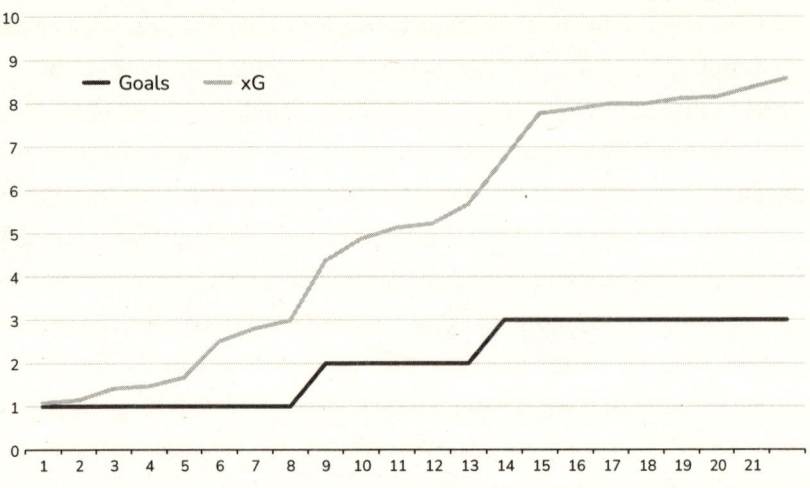

Figure 11.3: Kevin De Bruyne – Goals vs xG, 2020/21 (First 22 Matches)

basically half a season almost matched Werner's deficit over the entirety of the 2020/21 campaign.

At this point in the campaign, it might have been tempting to write De Bruyne off as a wayward finisher. Certainly, if he was a youngster starting out at the beginning of his career he would've been dubbed 'wasteful' and scolded for his lack of 'clinicalness'. Such a youngster might not have been given a chance in the professional game and ended up becoming a postman, estate agent or reality TV star. De Bruyne's track record meant he continued to be named in starting line-ups by Pep Guardiola, and what happened over the next year and a half made any doubts about his finishing ability look very silly indeed. In the 33 Premier League matches he played from that point up until the end of the next season, De Bruyne scored 18 goals from just 7.36(xG), an overperformance of nearly 11 goals and an xG conversion rate better than any player in European football's big-five leagues. Figure 11.3 depicts De Bruyne's goals and xG trend for the entire two-season period from the beginning of 2020/21 to the end of 2021/22. You can see the Belgian's goal drought at the beginning of the first season, before his scoring rate bounced back with a vengeance despite the fact his xG trendline actually slowed down slightly.

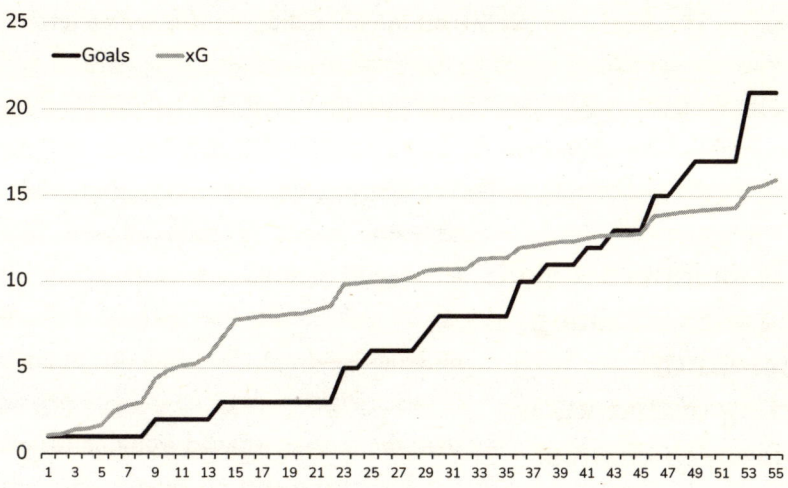

Figure 11.4: Kevin De Bruyne – Goals vs xG, 2020–2022

So, what does all this mean? Did De Bruyne take a shooting lesson before his 23rd appearance of the 2020/21 season? Did he change his kicking style, or perhaps invest in a new pair of boots? No, he didn't. De Bruyne was simply at the mercy of Lady Luck. The conclusion we should draw from De Bruyne's wild fluctuations in xG conversion, and that of countless other players, is that 'finishing ability' of players is prone to a large degree of variance, even over relatively long periods of time. The leading goalscorers will only amass around 100 shots in any given season. This is still a relatively small sample size and only 15 to 20 of these efforts will actually result in goals. All it takes is for a handful of attempts which otherwise would have hit the back of the net to actually strike the wrong side of the post or be stopped by an unusually athletic save and suddenly your goals tally is looking a lot worse off.

The Expected Goals Philosophy made the point that 'finishing ability' doesn't differ as much as you might expect it to. Cristiano Ronaldo is widely acknowledged as one of the world's greatest finishers, scoring more than 800 goals over the course of his career. However, he underperformed xG in six of the final eight seasons he spent competing in the top five European leagues. Despite scoring a whopping 186 goals over this period, his shooting opportunities amassed to 192.36(xG). What this means is Ronaldo actually scored fewer goals than the average striker would be expected to score given the chances he had. What set Ronaldo apart wasn't his ability to finish off chances, but his ability to obtain high-quality chances in the first place. His skill at dribbling past players, getting on the end of crosses, finding space in the box, and reading where the ball was going to drop allowed him to consistently churn out an extraordinarily high xG output. Studies have estimated that these qualities allowed Ronaldo to accumulate between eight and 12 (xG) more per season than the average forward would in his place.

There is an inherent skill involved with attaining xG, whereas the converting of that xG is more down to chance.

The notion that we should praise players for collecting xG rather than converting xG is counterintuitive. Consider Werner's attempt at the beginning of the chapter whereby the German followed in Mount's rebounded shot but put the sitter wide. It's normal for fans to berate Werner for such a miss, but think of the players who wouldn't have followed the shot in and therefore wouldn't have even had the opportunity to miss the chance. Ironically, Werner could have escaped criticism by simply being lazy and not gambling on the goalkeeper spilling the ball. There is an inherent skill involved with attaining xG, whereas the converting of that xG is more down to chance. Werner's reading of the game and the effort he made to follow in the shot should be recognised because these are the things in his control. Whether or not he finishes the chance is, in a sense, less important if we're trying to analyse how good Werner is, because he could be denied by a great goalkeeping save, the unfortunate bobble of the ball, or a number of other things out of his control.

Brentford discovered early on that a striker's ability to achieve a high xG output was more important than their xG over-performance or their supposed ability to finish chances. Over the years, the Bees have focused heavily on creating a large volume of high-quality chances. When they were a mid-table Championship team, they were always towards the top end of the goalscoring charts. Indeed, they were the joint seventh top scorers in the Premier League in 2022/23 despite having the lowest wage budget. Over the years, several strikers passed through the corridors of the West London side – Andre Gray, Scott

Hogan, Neal Maupay, Ollie Watkins, and Ivan Toney, to name but a few. Whichever number nine came through the doors, they were told simply to get in the danger zone around the six-yard box and second six-yard box, the positions of maximum opportunity. 'Your teammates will find you, just get yourself in there.' Ollie Watkins was perhaps the best at accessing these positions, scoring 25 goals for Brentford during the 2019/20 season, only one of which was struck from a position further back than the penalty spot. The other 24 all came from within either the six-yard box or the second six-yard box.

Brentford don't care much about 'finishing ability' because they know it doesn't matter as much as actually creating high-quality chances in the first place and getting in the areas which allow for the easiest goals to be scored. Sure, Neal Maupay might have been slightly less clinical than his predecessor, Scott Hogan, when the Frenchman first signed for the club in 2017. But Maupay was much better at getting around the six-yard box and scoring tap-ins, which is why he achieved greater success in the Championship at Brentford and garnered a move to a Premier League team (the xG-driven Brighton & Hove Albion, no less). Each of the five Brentford strikers listed above were sold for several times their initial fee – except Ivan Toney, who is now an England international still playing for Brentford at the time of writing and is currently valued at more than any of the others.

Finishing ability is a lot more standardised than most people believe. Sure, some strikers such as Harry Kane, Son Heung-Min and, of course, Lionel Messi do a possess 'clinicalness' above the norm and consistently score more goals than xG says they should. But these exceptional finishers are few and far between, and certainly account for a tiny portion of all shots taken across Europe's top five leagues. Most players act like random number generators when it comes to their goals output relative to xG.

Many fall victim to the hot hand fallacy when attempting to explain the various dry spells and hot streaks experienced by players like Kevin De Bruyne and Timo Werner. The hot hand fallacy is a cognitive bias which leads people to believe that a person who has experienced a series of successes is more likely to succeed again in the future. In other words, it's the belief that success breeds success. For example, a basketball player who has made several consecutive shots may be believed to have a 'hot hand' and to be more likely to make the next shot. However, research has shown that the hot hand fallacy is just that – a fallacy. The probability of a basketball player making a shot does not increase after a successful shot; each shot is independent of the one that came before it and the player's success is determined primarily by their skill level and the circumstances of each shot. Despite substantial evidence, many continue to believe in the hot hand fallacy. This misguided faith is likely driven by a desire to find patterns and meaning in randomness, as well as by a yearning to believe that success is more predictable and controllable than it actually is.

Study the cumulative goals versus xG charts of the four strikers in Figure 11.5. The straight line represents the cumulative xG of a player, while the zig-zagging line represents the number of goals they scored. Certain narratives will immediately come to mind about each footballer. Player A's finishing was generally consistent with what would be expected of him. Player B performed below par and suffered two long goal droughts where he seems to have lost his way. Player C had a terrific run of form, particularly towards the back end of this period of matches. You can almost sense the confidence running through his

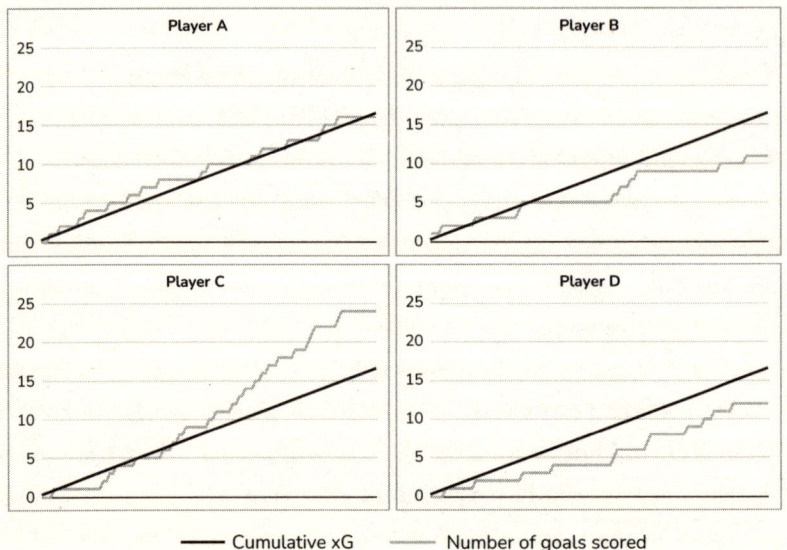

Figure 11.5: Goals and xG Performances of Four 'Players'

veins as he scored goal after goal, defying the xG figures. Player D suffered a similar spell to Player B and is probably also a poor finisher.

Now, what if I told you each of these charts was generated completely at random? Each chart features 100 'shots' which were in fact rolls of a normal six-sided dice. Every time the dice landed on a 'six,' it counted as a 'goal.' The 'xG' line in fact shows the cumulative expected value figure given that each roll possessed 0.1667(xG) – a one in six chance of the 'shot' (roll) resulting in a 'goal' (six). The 100 rolls overall mean each 'player' accumulated 16.67(xG) in total.

As a side note, some might question whether the above charts were actually generated at random. Indeed, they don't look particularly random. For those doubters, consider the story of a professor who used to ask his classes of students to produce four charts, each with a sequence of 100 coin flips similar to the charts outlined above. Heads should be +1 on the line graph and tails should be -1. Three of the charts should be generated randomly by tossing coins and plotting the results on the chart. The other one should be generated by the students, who should try to make it look as random as possible. The professor left the room and when he came back the students had the four charts waiting for him. He could almost always immediately work out which one had been artificially generated by the students, because the one designed to look random always looked *too random*. When flipping a coin 100 times, there are regular sequences when the coin lands on either heads or tails four or five times in row. You can test this using your own coin and a pen and paper (or by finding a coin flipping generator online). But in trying to make the coin flips look random, the class deviate between heads and tails too often in their self-generated chart and rarely introduce long sequences of one or the other. Random charts often don't look random, and vice versa.

Back to our mischievous xG example above. Because the scoring outputs of the 'players' were generated randomly, they were all just as likely to overperform as they were to underperform over the course of the whole season. Not only were each player's shots independent of one another, but all four players actually possess the exact same level of 'ability' or 'skill.' If you ran Player C's simulation again, they would have the same chance of succeeding as any of the other players. This example shows how easily our brains attach meaning to randomness.

Pundits and spectators might mistakenly claim Player C had a 'hot hand' (or should that be 'hot foot'?) in the second half of this period when in fact this performance was the product of nothing but pure randomness.[14]

Forming narratives around such performances is as pointless as trying to identify patterns in lottery numbers. Some people might look for numbers which haven't come up in a while, whereas others might choose the dates of their loved ones' birthdays. But neither of these factors make the chosen numbers more likely to come up in the next draw. The machine doesn't deal in sentiment, it deals in randomness. Football is not a lottery, and not everything in football is chance. But it is a lot more at the whim of random occurrence than we are led to believe, particularly when it comes to finishing ability.

Recognising and overcoming the hot hand fallacy is important for making rational decisions in many areas of life, including sports, finance, and gambling. Consider a professional bettor who tends to make bets which he estimates all have a 50 per cent chance of paying off. Even if they've found an edge, the bettor is still likely to go through hot streaks and cold streaks just like the simulated players. The same is true for football, where the better team only wins around 55-60 per cent of the time. We read a lot into the form of teams – how many recent games they've won or lost. But these results are as good as meaningless over a short sample and can even be misleading over the course of half a season or so. Saying a team has won their last five matches is often as good as saying a coin has come up heads on its last five flips. Form charts often trick us into drawing narratives where there aren't any. Form certainly exists in some guise, but there's a lot more randomness at play than meets the eye.

Matthew Benham once reflected on the difference between England and Germany's penalty shoot-out record in major tournaments. Before the World Cup in Russia, England had lost six times in seven shoot-outs, while the Germans had won all but one shoot-out in their history. Various studies sought to explain the reasons for these sequences, citing football education or national character.

---

[14] The probability of one of these 'players' scoring 24 goals or more – as Player C managed – is about one in eight. Given we produced four of these charts, the likelihood of one 'player' scoring at least that many was fairly substantial.

But, as Benham outlined, 'If football matches were decided by a coin toss rather than a shoot-out, like they were in the old days, we might still have seen a similar series of results. The odds of England defeating Germany in their next shoot-out against one another would likely be close to 50-50.'

Now consider Figure 11.6. This graphic features the exact same scoring pattern of each of the previous players, but the supposed xG output has been altered. Put yourselves in the shoes of a Head of Recruitment at a football club. You're tasked with signing a new striker for your team and have been presented with these four profiles. Which one would you plump for? All other things being equal, Player A and Player B can probably be discarded fairly quickly as neither the goals tally nor the xG generation is particularly exciting. The question then becomes who is preferable between Player C and Player D. What we know for sure is that the market will place a much greater value on Player C. Goals are the ultimate currency in football, and Player C scored 24 of them from their previous 100 shots. A large transfer fee and wage packet would be required to secure their services.

On the other hand, Player D only found the back of the net on 12 occasions but was able to get into much better scoring positions, accumulating 20.00(xG)

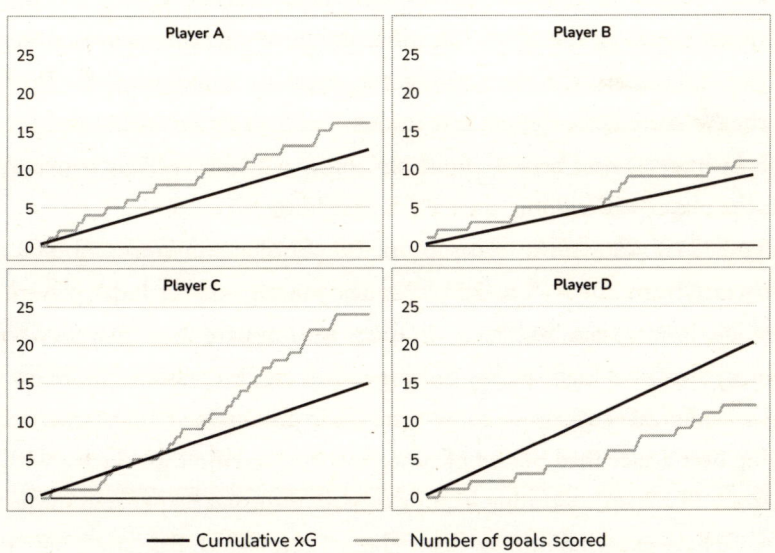

Figure 11.6: Different Goals and xG Performances of Four 'Players'

to Player C's 15.00(xG). The ultimate question is: would you rather sign a striker with a large goals output but low xG, or large xG output but fewer goals? This cuts right to the core of the debate surrounding finishing ability.

If you were to ask Liverpool, the answer would be straightforward. John Henry's xG-oriented recruitment team signed Roberto Firmino from Hoffenheim and Sadio Mané from Southampton on the back of seasons where they didn't set the world alight in terms of goals but were producing strong xG figures. Both players were towards the top of the global list of realistic signings for xG+xA. The top player in this sample at the time? Mohamed Salah, the AS Roma forward who was undervalued in the market because of doubts over his ability to succeed in England, having previously been discarded by Chelsea. Liverpool signed these three players and they became one of the most prolific attacking trios in European football, notching over 300 goals for the Reds. They were bought for less than a combined £120 million but, at their peaks, reached a valuation of more than £470 million.

A few seasons later, Liverpool repeated the feat by signing Diogo Jota, who had scored seven goals from 12.08(xG) for Wolves in 2019/20. Liverpool signed him for a mere £40 million and the Portuguese went on to score 24 league goals from 24.43(xG) in his first two seasons at the club. You can employ the brightest recruitment minds in the sport, but often the most effective transfer strategy is simply to sign players who are accumulating excessive amounts of xG. The fewer goals they've scored, the better, as it means they'll likely be undervalued by the rest of the market. And because finishing ability will always likely regress to the mean, the player will 'cash in' their xG before long.

Figure 11.7 shows the most proficient finishers in terms of xG over-performance from 2014/15 to 2021/22. The sample only includes players who accumulated more than 50.00(xG) in order to eliminate those players who had brief lucky streaks of high-quality finishing. This chart represents the crème de la crème of 'clinical' strikers – those who have maintained high standards of finishing over a sustained period of time. Son Heung-min tops the list with 104 goals from 75.21(xG), meaning he hit the back of the net 38 per cent more often than you'd have expected him to. Eden Hazard and Gareth Bale come next in the rankings. The one-time Real Madrid colleagues might seem surprisingly high up,

Player	Goals	xG	Goals / xG	Pens Scored	Pens Missed	Pen %
Heung-Min Son	104	75.21	138%	7	5	58%
Eden Hazard	65	52.22	124%	49	8	86%
Gareth Bale	77	64.43	120%	12	3	80%
Antoine Griezmann	119	99.69	119%	19	11	63%
Lionel Messi	237	202.44	117%	109	31	78%
Paulo Dybala	95	81.40	117%	28	5	85%
Kylian Mbappe	136	117.75	115%	25	6	81%
Harry Kane	180	159.19	113%	60	11	85%
Ciro Immobile	160	143.32	112%	69	16	81%
Mohamed Salah	153	140.53	109%	30	6	83%
Total	1326	1136.18	117%	408	102	80%

Figure 11.7: Biggest xG Overachievers (2014–2022)

but they both possess a smaller sample of shots and are therefore likely to deviate more aggressively from the norm.

Suppose you present an average striker with enough chances that he scores 100 goals, how many goals do you think an elite, world class striker would score from those exact same opportunities? It's not uncommon for answers to range from 140-150. The chart above shows that the best finishers will only score about 117 goals. In other words, they score goals at a rate only 17 per cent faster than the average striker. The ability to convert chances at this rate certainly shouldn't be sniffed at and can make a difference over the course of several seasons. But the impact of 'elite' finishing ability probably isn't as big as you thought it was. It's also worth considering the inconsistency in scoring rates, even among world-class footballers. Every single player in the list has underperformed xG over the course of at least one season in their careers. Consider the case of Hazard, who had outperformed xG in all five of his final seasons at Chelsea. When Real Madrid signed him, they thought they'd signed one of the most clinical players on the planet, but when he arrived at the Bernabeu, the Belgian underperformed xG in three of his first four seasons and scored just three goals from 5.85(xG) overall. No player's 'clinicalness' can resist the gravitational force of xG forever.

We can further test the existence of finishing ability by looking at the penalty success rate of Europe's most clinical players. The right-hand columns on Figure 11.7 show the all-time penalty records of each player. Penalties can be considered the ultimate test of finishing ability; they are an exercise in clinicalness condensed into a single action. A penalty represents the easiest, most practicable and repeatable scoring opportunity for a player in any given game. The leading xG models rate penalty kicks at between 0.77 and 0.80 (xG).[15] Summing up the overall penalty record of this cohort of the world's strongest finishers shows an 80 per cent success rate with 408 out of 510 penalties being scored, roughly in line with what the average player would expect to achieve. This finding somewhat undermines the notion that finishing or clinicalness is an innate or repeatable skill.

WHEN WHOLE SQUADS DEVIATE FROM EXPECTATION

There is a fundamental difference between how we assess the xG overperformance or underperformance of players versus how we assess that of teams. Many still believe, contrary to the argument laid out above, that 'clinicalness' is an innate and repeatable skill which lots of players possess and lots of others don't. This belief mechanism lays the blame with the player when he underperforms xG over a certain period, rather than with the invisible hand of chance. When Kevin De Bruyne went through his dry patch, it was the fault of the Belgian rather than the result of the natural variations in goal outputs. According to these people, De Bruyne is not a random number generator but rather a fully autonomous machine. But even the most fervent subscribers to this school of thinking will take a different stance when it comes to the xG underachieving of entire teams. Whereas the xG underperformance of players is usually put down to skill, the xG underperformance of teams is put down to bad luck. In teams, the blame is shared across a number of players, so it's easier to accept that deviations from the

---

[15] The difference between models results from the fact that each model is trained using a different set of data. Note that a penalty will be the same xG for every spot-kick within that particular model – in other words, a model that rates a penalty as 0.79(xG) will rate every other penalty at 0.79(xG) as well.

norm are a product of chance. There is a diverse portfolio of 'finishing ability' which should offset the poor quality of any single player.

Consider Brighton's catastrophic 2020/21 season where they scored 39 goals (excluding own goals) from 55.06(xG). The Seagulls scored 16 goals fewer than they would have been expected to given the quality of chances they produced. A closer look at the xG performances of each player in their squad shows the main culprit was Neal Maupay, who scored eight goals from 13.79(xG). This is certainly a significant xG underperformance – the Frenchman scored around six fewer than he should have – but if you take him out of the

Good players in good teams will accumulate high xG figures, while bad players in bad teams will accumulate low xG figures.

equation then Brighton still scored 10 fewer than expected. What's the makeup of these invisible 10 goals? Who should have been scoring, but didn't? Well, the answer is almost everyone. No player other than Maupay underperformed by more than 2.50(xG), while an astonishing 18 of the 22 Brighton players to take a shot that season scored fewer goals than expected.

Again, think of each squad of players as a group of random number generators. Each of the team's players collects an xG figure over the course of a season based on the skill of both himself and the wider team. Good players in good teams will accumulate high xG figures, while bad players in bad teams will accumulate low xG figures. Think of all the squads in Europe and all the recent seasons they've played. In a standard season, you'd expect roughly half the players to underperform their xG and roughly half to overperform. Across the squad, you'd expect the number of goals to generally align with the total xG accumulated. Now think of the possibility of one of these hundreds of squads having a catastrophic season whereby almost all of their players underperform. Nearly all the random number generators come out with less-than-favourable totals. That's what happened to Brighton in 2020/21.

Now think of all the shots a team concedes over the course of a campaign. This collection of attempts is also being produced by random number generators, but these ones are playing for opposition teams. Suppose this cohort of footballers happens to overperform xG. A few long-range efforts dip under the crossbar and into the net when they otherwise might've whistled over. A few blocked shots end up deflecting into the bottom corner of the goal rather than out for a corner kick. That's also what happened to Brighton that season, as they conceded 46 goals from 41.43(xG). A near-impossible combination of their own players underperforming xG and opposition players overperforming xG led to one of the unluckiest campaigns in modern history.

It's pure coincidence that some of the most unexpected xG seasons in recent history have been experienced by teams who arrived early to the Expected Goals party. Brighton endured their infamous 2020/21 campaign, while Brentford's first Premier League season saw them collect 10 fewer points than expected given their xG performances. Perhaps the most significant xG *overperformance* in modern times was enjoyed by the third member of this analytical triad: Liverpool.

By almost every conceivable metric, Liverpool didn't deserve to win the Premier League title in 2019/20. They accumulated a goal difference of 52 compared to Manchester City's 67. If you think of the season as a single 3,420-minute-long game of football, Liverpool won with a scoreline of 85-33 while City won 102-35. The Reds also won eight of their opening 14 matches by a one-goal margin, scoring a handful of late winners in the process. These key indicators implied that Liverpool weren't as good as the league table suggested.

If these stats hint at an element of good fortune, the xG data confirms it. Liverpool scored roughly 10 goals more than expected and conceded roughly seven fewer than expected, meaning their goal difference of 52 should've been more like a goal difference of 35. Manchester City's xG difference was almost twice as high as this, at 65. Liverpool's performances over the campaign merited 74 expected points, 13 fewer than City's 87 expected points total and only one greater than Chelsea's tally of 73(xP). According to xG, Liverpool should have finished 13 points behind City, when in reality they finished 18 points above the Citizens. The combination of Liverpool's overperformance and City's underperformance relative to xG meant this was a one-in-a-100 season. But the thing about seasons is, hundreds of them take place across Europe every decade.

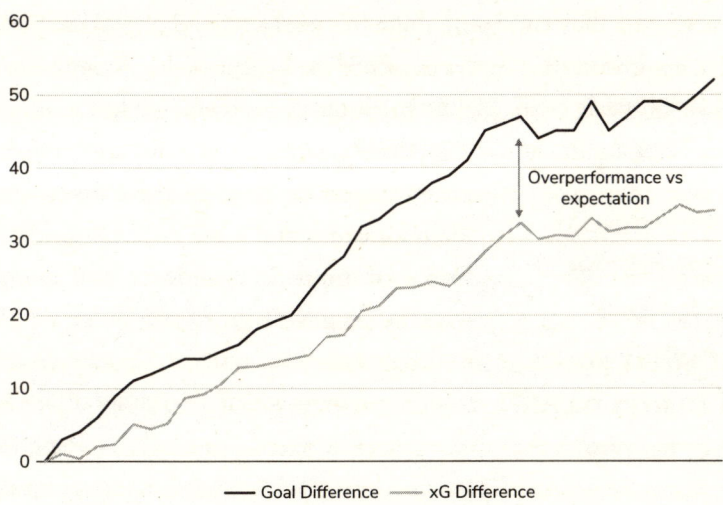

Figure 11.8: Goal Difference vs xG Difference, Liverpool 2019/20

How can we explain Liverpool's overperformance? An immediate temptation is to credit the supreme finishing ability of their players. The casual supporter might think of the likes of Mohamed Salah, Sadio Mané and Roberto Firmino as notoriously ruthless finishers, highly skilled at finding the back of the net. Surely they're always likely to outperform xG? Well, this isn't quite the case. For one thing, Salah and Firmino actually *under-performed* their xG in 2019/20, scoring 19 goals from 20.44(xG) and nine goals from 16.87(xG) respectively. This is crazy: in a team that scored 10 goals more than expected, two of their three primary attacking players actually scored around nine fewer goals than expected between them. That means the remaining players scored around 19 goals more than xG dictates they should have done. What's more, not a single individual Liverpool player overperformed by more than 3.50(xG). It turns out that Liverpool's overperformance came as a result of the exact opposite reason for Brighton's underperformance in 2020/21: almost every player in their squad scored slightly more goals than expected. The likes of Trent Alexander-Arnold, Virgil Van Dijk, Georginio Wijnaldum, Alex Oxlade-Chamberlain, Divock Origi, Jordan Henderson and Fabinho all chipped in with two or three goals more than they should have done given the quality of their chances. The random number generators happened to come up trumps for the Reds.

Some sceptics didn't accept that the role of chance had played such a big part in delivering Liverpool's first ever Premier League title. Among the more ridiculous rebuttals was the claim that Jürgen Klopp's men had found a way to 'hack' xG. These arguments claimed that Expected Goals didn't apply to Liverpool for one reason or another. There were claims that the Reds had found a way to cheat game state, or that their supreme confidence and momentum was trumping the logic of xG. We've already seen in this chapter how easily we can attach meaning to performances which are simply driven by rolls of a dice. The proof was in the pudding the next season; Liverpool scored four fewer goals than their xG merited, finished on 69 points and 69 expected points, and scraped into the Champions League places on the final day of the campaign. The media narrative in 2020/21 was about how far Liverpool had dropped off as they amassed 30 points fewer than the prior campaign. In reality, though, they were only five expected points worse off. Their performances had remained fairly consistent, but the extreme fortune they had gained in 2019/20 had abandoned them. Lady Luck doesn't tend to hang around for long.

Look beyond actual results, beyond the league table, and instead at what you deserved to win based on your performances.

Timing is everything in sport. Elite golfers in the era of Tiger Woods might wonder what their trophy cabinet would look like if they hadn't been playing at the same time as one of the all-time greats. Top-class tennis players competing in the same period as Roger Federer might reflect on how much silverware they would have collected if it weren't for the misfortune of competing against the Swiss legend. In business, too, Blackberry executives might ponder on what might have been had the iPhone not arrived on the scene, while Yahoo might curse their luck that Google came to be. Liverpool may look back in a similar way on the competition between themselves and Manchester City between 2018 and 2022. The Reds finished on 97 points in 2018/19, a total which would have won them the title in 25 of the 26 prior Premier League seasons, but lost out on the title by one point. Liverpool pushed the Citizens all the way again in 2021/22, reaching 92 points but once again losing out to City by a single point. In such situations it can be useful not to reflect on the final outcome of the season or how much silverware was won, but rather on the performance of the team. It sounds strange, but resisting the temptation to compare yourself to your opponents and choosing instead to think about and reflect on the things that were in your control is the best course of action. Look beyond actual results, beyond the league table, and instead at what you deserved to win based on your performances.

Expected Goals allows you to look at the true performances of either team in each game. This insight can be extrapolated over each season or, indeed, a number of seasons to reveal how well a team truly performed over a given period of time. The one-season case study of Liverpool's title-winning season certainly reveals a campaign blessed with the backing of the xG gods. But over the course of Klopp's reign, does one Premier League title seem like a fair reward for the fruits of his labour? We can use the expected points totals Liverpool have accrued over Klopp's time in charge to work out how many titles you'd expect him to have won based on his team's xG performances.

Figure 11.9 shows that Liverpool didn't offer any meaningful title challenge in Klopp's first two years at the club. Their first significant push came when

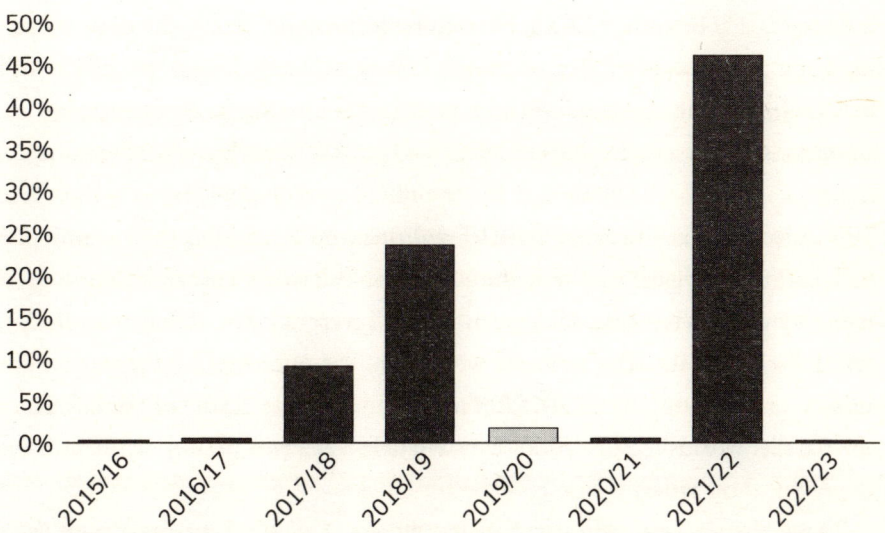

Figure 11.9: Chance of Liverpool Winning the Premier League Based on their xG Performances

they collected 79 expected points in 2017/18, meaning they could have expected a 9 per cent chance of winning the league. In other words, if they performed at exactly the same level each and every season, creating and conceding the exact same amount of xG in each game, they could expect a 9 per cent chance of lifting the trophy in any given campaign. They improved again in 2018/19, taking their chance of league victory up to about one-in-four. The year they finally won the title in 2019/20 was actually one of the season's they deserved it least, performing well enough to win the league just one-in-50 times. But one-in-50 occurrences have a habit of occurring one-in-50 times. The 2020/21 campaign was Liverpool's worst since Klopp's first season, but they bounced back in 2021/22 with a season's worth of performances good enough to win the league 46 per cent of the time. 2022/23 went poorly both from an Expected Goals and actual goals point of view, as the club missed out on a top-four finish. Overall, Liverpool performed well enough to earn 0.82 'expected titles' over this eight-year period – not too dissimilar from the one title they actually earned.

Further analysis can tell us the likelihood of Klopp winning a certain number of titles over his reign. Imagine 100 different parallel universes where Klopp's Liverpool produce the exact same xG figures in every game they played between 2015/16 and 2022/23. Everything is the same, except the goals output the team produces is subject to change. Every Mohamed Salah penalty is still 0.77(xG), but the outcome of those penalties is unwritten. In essence, we are simulating the first eight years of Klopp's reign 100 times but with the same xG figures. Figure 11.10 shows that Liverpool fail to win any titles in 44 of these 100 universes. Fans in these parallel realities would bemoan their team's bad luck and think, rightly so, that the Reds should have at least one title to their name by now. What this does mean is that Liverpool win at least one title in 56 of the universes. The universe we live in, the one where Liverpool's eight seasons under Klopp up to 2022/23 produced one title, is one of 36. Fifteen of the parallel worlds see Liverpool win two titles over this period, while four see them win three titles.

And there's one solitary universe where Klopp's Liverpool won four Premier League titles in his first eight years at the club. Imagine what life in

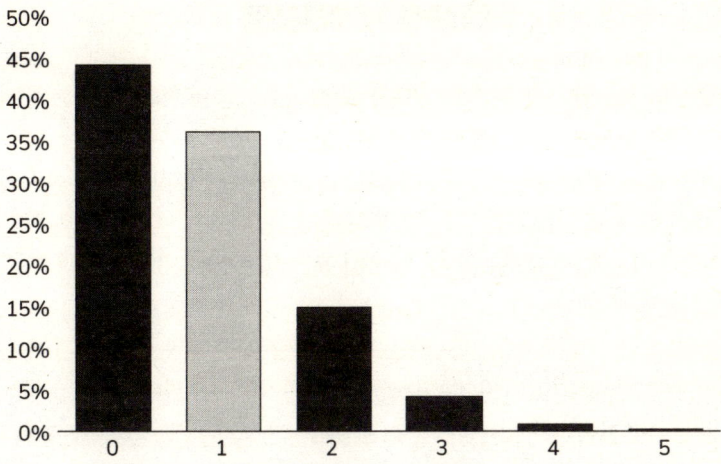

Figure 11.10: How Many Titles Should Jürgen Klopp Have Won at Liverpool?

this universe is like. Klopp must be hailed as the greatest manager of all time. Mohamed Salah probably has three or four Ballon d'Ors, while Manchester City fans are likely bemoaning the incredible bad luck they've suffered to only have won a couple of Premier League titles under Pep Guardiola. Of course, opposition fans would also dub Liverpool as one of the luckiest teams in modern history.

Remember, the quality of their performances in this universe is exactly the same as the one in our actual universe – the only difference is in the conversion of those chances at either end. In the world where Liverpool are four-times Premier League champions, they must have finished their chances at an extremely high rate and opposition players must have missed an unbelievable number of sitters.

Analysis of this kind can be carried out for any team and helps provide a sense of perspective. It can be difficult not to be swept up in the emotion of football, whether that be winning the league or being relegated. Looking at what would have been expected to happen based on performances can provide objective clarity. Liverpool have deserved to win a league title under Klopp, it just wasn't delivered in the season you'd expect. One thing is for sure: it might have panned out differently if they'd been successful in their attempts to sign Timo Werner.

CHAPTER SUMMARY

- Expected Goals tells us how many goals a player should have scored from the chances they were presented with.
- When strikers massively outperform or underperform their xG, is it down to skill or luck? The answer is probably a combination of both, but historically we've overvalued the role of skill and underplayed the role luck plays in 'clinicalness'.
- Players who amass large xG totals are probably more valuable than those who have outperformed xG over a certain time period.
- Looking at the make-up of a squad of players' xG performances can give crucial insights into how lucky they were in any given season, and can help us determine how well they actually played over a series of seasons.
- Expected Goals allows football to be viewed more analytically and objectively, removing emotion from the experience, as it offers a clarity as to what was expected to happen based on performances, rather than what did happen.

12

THE ADVANCED ANALYTICS TOOLKIT

BEYOND THE REALM OF XG

*'If you challenge conventional wisdom,
you will find ways to do things much
better than they are currently done'*

Bill James, American baseball historian and inventor of Sabermetrics

Expected Goals has become the lead singer in the band known as Football Analytics. It has caught the attention of the wider public and performs in front of a (mostly) adoring audience each week. But no band is comprised of just one musician; there are several other members of the group all contributing to the broader movement. Indeed, there are some songs which traditional xG simply can't sing. Either the notes are too high or the tempo is too quick. In these cases, there is usually a back-up singer waiting in the wings ready to take centre-stage.

Traditional xG isn't perfect. We have already studied how game state can skew Expected Goals stats. Traditional xG also doesn't tell us a great deal about what happens after a shot takes place because it only provides a snapshot of what is happening in the instant the player strikes the ball. Following the precise moment of the shot, we only know whether the effort ended in a goal or not. This prevents us from gleaning much insight into how good the attempt actually was, which restricts our analysis of the shooting proficiency of forwards and the save-making ability of goalkeepers. For example, consider a striker who takes a shot from outside the box which curls dangerously towards the top corner, only for the goalkeeper to acrobatically tip the effort over the bar. Standard xG will show the striker scored 0 goals from, say, 0.03(xG). This is useful information but doesn't capture a true picture of what took place. The data should ideally reward the striker for the high-quality attempt, while the goalkeeper should be credited for pulling off a fine save.

The antidote to this problem is known as 'Expected Goals on Target,' or xGOT. What xGOT seeks to measure is the likelihood of an on-target shot resulting in a goal based on the combination of the underlying chance quality (xG) and the end location of the shot within the goalmouth. Shots which end up in the corners of the goal will be given more credit than those which end up down the goalkeeper's throat. More advanced xGOT models will also consider factors like the trajectory and power of the shot, both of which influence how likely it is to beat the goalkeeper. Just like xG, xGOT is measured on a scale of 0.00 to 1.00, whereby an attempt which the goalkeeper should save every time (or isn't on target) merits 0.00 and an attempt which no goalkeeper in the world would save would theoretically merit 1.00.[16]

Most football fans will remember Olivier Giroud's scorpion kick against Crystal Palace in January 2017. A fast Arsenal break left Alexis Sanchez on the left wing in space. The Chilean delivered a cross into the box towards an onrushing Giroud, but the ball was played slightly behind him. Giroud skilfully

---

[16] In fact, a shot on target will never have a 100 per cent or 0 per cent chance of going in, just as we can't be 100 per cent certain that the sun will rise tomorrow or 0 per cent certain that Elvis Presley won't be found water skiing across Loch Ness on the back of a monster.

adjusted his body and 'scorpion kicked' the ball in off the underside of the crossbar from 12 yards out. The goal, scored on the first day of 2017, ended up winning the FIFA Puskás Award for the best goal that year. The shot merited 0.05(xG), implying a chance of this nature would be scored one in every 20 times it was attempted.[17] The post-shot model would look at where the shot ended up – in the top corner deflecting in off the crossbar, an incredibly hard effort to save for Crystal Palace's Wayne Hennessey – as well as the power and trajectory of the strike, and assign the attempt something like 0.86(xGOT). In other words, Olivier Giroud managed to turn a shot which the average player should score 5 per cent of the time into a shot which would beat the goalkeeper 86 per cent of the time.

It might already be becoming clear how xGOT can help analyse shooting performance. A player whose xGOT consistently exceeds their xG is executing their shots well. The difference between the xGOT and xG figures of a shot or series of shots is known as 'shooting goals added' (or SGA). Figure 12.1 shows the 'shooting goals added' of the top ranked xG strikers of the 2022/23 Premier League season. Harry Kane scored 30 goals from 23.15(xG), meaning he bagged around seven goals more than expected. But how much of that overperformance was a result of his shot execution? Kane registered 29.35(xGOT) that campaign, meaning his 'shooting goals added' was 6.20. Compare this to Erling Haaland, whose shot placement only led to 1.11 SGA over the same season. Despite scoring more goals than Kane, you could argue that Haaland's execution wasn't as good because he didn't convert his xG into xGOT as well as the Englishman.

Note that a strong xG to xGOT conversion doesn't necessarily imply innate 'finishing ability'. Players can go through hot streaks of producing great xGOT figures, before suddenly going through a dry patch where their xGOT output drops off. Just as xG conversion can frequently be put down to little more than chance, so too can xGOT overperformance or underperformance. For instance,

---

[17] In reality, the effort was so unusually skillful that it probably merits an xG lower than this. A minor flaw with xG is that it doesn't generally recognise attempts such as scorpion kicks, bicycle kicks, rabonas and the like. This doesn't tend to cause big problems as these types of shots are rare and the difference will only be a few percentage points. For the sake of ease, we'll stick to the figure of 0.05(xG) for Olivier Giroud's attempt.

Player	Club	Goals	xG	xGOT	Shooting Goals Added
Erling Haaland	Man City	36	32.91	34.02	1.11
Mohamed Salah	Liverpool	19	23.27	26.25	2.98
Harry Kane	Tottenham	30	23.15	29.35	6.20
Ivan Toney	Brentford	20	21.40	23.72	2.32
Callum Wilson	Newcastle	18	18.67	16.76	-1.91
Marcus Rashford	Man Utd	17	17.34	21.64	4.30
Ollie Watkins	Aston Villa	15	17.09	16.37	-0.72
Aleksandar Mitrovic	Fulham	14	15.67	16.57	0.90
Gabriel Jesus	Arsenal	11	16.03	13.67	-2.36
Darwin Nunez	Liverpool	9	14.41	10.55	-3.86

Figure 12.1: Shooting Goals Added, Premier League 2022/23

Haaland hit the woodwork five times during the 2022/23 campaign, while Kane only hit it twice. Suppose three of Haaland's woodwork-hits had happened to fall the right side of the post and three of Kane's top corner strikes had happened to hit the crossbar. Each one of Haaland's shots would then become roughly 0.95(xGOT) instead of 0.00(xGOT) and vice versa for Kane. Suddenly their respective xGOT outputs look very different.

ANALYSING GOALKEEPERS

Historically, the analysis community has struggled to accurately evaluate the ability of goalkeepers. Clean sheets and save percentages are perhaps the most popular way of determining the quality of shot-stoppers. Indeed, the Golden Glove award is presented to the goalkeeper who keeps the cleanest sheets each season in the Premier League. But these metrics have limitations. A terrible goalkeeper who is playing behind the best defence in the world will likely keep a large number of shutouts. Meanwhile a poor goalkeeper playing behind a defence which limits its opponents to long shots will likely have a high save percentage, but with many of these saves proving extremely comfortable.

Measuring goalkeepers using traditional xG, by looking at the difference between the xG a goalkeeper faces and the number of goals they concede, is also problematic because it counts off-target shots and doesn't speak to the quality of on-target attempts. A goalkeeper might face shots worth 9.99(xG) and not concede a goal, but if none of these shots actually hit the target then we know nothing about the ability of the keeper. This is where xGOT comes in. Given the goalkeeper is the only thing standing between an on-target shot and a goal, it has proven incredibly useful in demonstrating shot-stopping ability. A shot straight down the middle of the goal from the edge of the box may carry 0.05(xGOT), meaning it has a 5 per cent chance of going in but also, crucially, a 95 per cent chance of being saved by the goalkeeper. A goalkeeper would be expected to save an attempt of this nature 19 times out of 20. Summing the xGOT of all the shots a goalkeeper faces means we can assess how many goals they would have been expected to concede based on the quality of the shots they faced. It allows us to directly credit goalkeepers for their ability to prevent goals, irrespective of their team's defensive strengths.

Figure 12.2 shows the top five and bottom five goalkeepers in terms of 'goals prevented' in the 2022/23 Premier League season. Allison leads the rankings. Based on the quality of the shots on target that the Liverpool keeper faced, the average goalkeeper would have been expected to concede between 51 and 52 goals. Given that he only conceded 40 goals (excluding own goals), we can credit the Brazilian with preventing nearly 12 goals with his saves. Meanwhile at the other end of the rankings, Gavin Bazunu conceded an incredible 16 more goals than he should have done. Bazunu's poor performances were perhaps the primary reason Southampton were relegated in last place.

José Sá also ranks poorly in 2022/23, conceding around six more goals than the average goalkeeper would be expected to. This is interesting because he ranked as the best goalkeeper in the league the previous season with a 'goals prevented' score of nearly seven. The case was made earlier in this book that the finishing ability of strikers is inherently random and unrepeatable. Perhaps this is also the case of goalkeepers' saving ability. Remember, xGOT only accounts for on-target shots which means it's a pure, distilled representation of a

Rank	Player	Club	Goals Against	xGOT Against	Goals Prevented	GP Rate
1	Allison	Liverpool	40	51.64	11.64	1.29
2	Bernd Leno	Fulham	49	58.25	9.25	1.19
3	David Raya	Brentford	43	50.19	7.19	1.17
4	Jordan Pickford	Everton	57	61.76	4.76	1.08
5	Emiliano Martínez	Aston Villa	34	37.62	3.62	1.11
21	Hugo Lloris	Tottenham	38	34.10	-3.90	0.90
22	Danny Ward	Leicester	41	36.93	-4.07	0.90
23	José Sá	Wolves	54	48.27	-5.73	0.89
24	Illan Meslier	Leeds	65	53.53	-11.47	0.82
25	Gavin Bazunu	Southampton	52	36.05	-15.95	0.69

Figure 12.2: Goals Prevented, Premier League 2022/23

goalkeeper's ability to keep the ball out of the net. That José Sá should perform so well one year then drop off the next hints at the impact of randomness. The same goalkeeper can produce vastly different performances from one campaign to the next.

'Goals prevented' is an intuitive measure of goalkeeper performance, but the inevitable rebuttal here is that goalkeepers who face more shots have the opportunity to 'prevent' more goals. Jordan Pickford faced 181 shots compared to Hugo Lloris' 117 during 2022/23, meaning he had 64 more chances to prevent a goal. This isn't necessarily a benefit: he also had 64 more chances to let a goal in. To allow for a fair comparison between keepers who face different shot volumes, we can look at their 'goals prevented rate.' Goals prevented rate is the number of goals that a goalkeeper was expected to concede as a proportion of the number of goals they actually conceded. For example, Emiliano Martinez possessed a better goals-prevented rate (1.11) than Jordan Pickford (1.08) despite the Everton keeper 'preventing' more goals overall (4.76). Normalising for the volume of shots allows us to see that Martínez expected to concede 1.11 goals for every goal he actually conceded, potentially showing him as a better shot-stopper than Pickford.

Expected Goals On Target allows us to isolate goalkeeper performance from that of their team in a way conventional metrics can't. The number and quality of shots a goalkeeper faces is still a product of the strength of the defence in front of him, but we are now able to see who is the most effective at preventing these shots becoming goals. In a way, we all have an internal xGOT model inside our head. In 2011, when Wayne Rooney scored a famous bicycle kick against Manchester City which arrowed into the top corner, we immediately recognised it as a shot that no goalkeeper in the world would save. The xGOT statistic is simply a mathematical representation of this feeling based on advanced modelling of the power and trajectory of shots.

Expected Goals On Target is clearly an incredibly useful metric for assessing the goalscoring ability of strikers and the goal-preventing ability of goalkeepers. But xGOT isn't flawless. For one thing, it's much more difficult to measure than xG. Traditional Expected Goals figures are taken primarily by looking at a shot's origin location, before building in factors such as whether it was taken on the stronger or weaker foot, the position of defenders, the type of pass assisting the shot, and so on. Each of these factors are fairly easy to determine. xGOT is slightly more technical because the trajectory and speed of a shot are tricky to accurately measure. Early steps are being made towards advanced ball-tracking capabilities which can better measure power and flight, but the accuracy of current methods is questionable.

There is also the problem of how xGOT is sampled. Constraining the sample to only on-target shots makes the construction of an accurate model more difficult. And any off-target shot or effort which fails to work the goalkeeper won't be registered, meaning blocked shots are excluded. Some players might count themselves unlucky that they executed the perfect shot but a defender happened to get in the way. Similarly, a striker who goes through an unlucky patch of hitting the post or crossbar will garner 0.00(xGOT) from these attempts, even though they were good efforts which came close to scoring. For instance, Frank Lampard was technically credited with 0.00(xGOT) for his 'ghost goal' against Germany at the 2010 World Cup. Despite these flaws, xGOT remains a valuable tool for telling us what happens after a shot takes place in a way that xG can't.

EXPECTED THREAT

The second main limitation of traditional xG is that it is only registered when a player takes a shot. Threatening attacks regularly occur which do not produce attempts at goal. How often have you seen a dangerous ball fired across the face of goal narrowly miss the outstretched foot of a striker? How many times have you seen a forward just beaten to the ball by a goalkeeper flying out of his box? How often have you seen an attacker round the keeper, only to be forced too wide to take a shot? These are all situations where a team has clearly created a chance to score, but the opportunity won't be reflected in xG data. Thus, a great deal of information about the chance-creating qualities of a team (and the chance-conceding tendencies of their opposition) is being missed out. Similarly, own goals are not recognised in traditional xG despite still representing a poor piece of defensive play or a good piece of attacking pressure. This should be reflected in the data, but isn't because there is no actual shot taking place.

Football analytics has found the solution to this problem in the form of another metric. Suppose you're at a football match and you fancy a half time Chicken Balti pie. You go to the concourse and see four lines all being served. The left-most queue is the shortest and seems to be serving the quickest, so you join the back of it. Once you've collected your steaming pastry and retaken your seat for the second half, you watch as your central midfielder ignores a sideways pass right and instead sprays a ball out to the winger high up the pitch on the left flank. You reflect on the similarity in decision-making between yourself and the footballer. When queueing for the pie, you made a mental choice to join the quickest line; the line which was most efficient and most valuable in terms of getting your pie as soon as possible. Meanwhile the midfielder chose the pass which he thought would be most likely to lead to a goal for his team. There's no specific name for the decision-making process you went through – perhaps xPies would be suitable – but football analysts have been able to model the value of various areas on a football pitch and use this to assess the chance-creating ability of players. The metric is called xThreat.

Think of a football pitch as a grid comprising multiple squares, sort of like a chess board. Which squares are the most valuable to have the ball in? Well, having possession just outside your own box isn't particularly useful if you're trying to score a goal. Having the ball midway inside your opponents' half is better. A shot from this position would register a low xG, but the chance of scoring from your current possession is relatively high, particularly if it's a moment of transition and the opposition defence is unset. Having the ball in your opponents' six-yard box is the best area if you're looking to score. The traditional suite of football metrics is great at telling us who is attacking the opposition goal through stats like goals, shots and xG, as well as telling us who is helping create the chances via stats like assists, chances created and xA, but it doesn't tell us a great deal about how the ball arrived at these dangerous positions in the first place. This is what xThreat seeks to do.

	1	2	3	4	5	6	7	8	9	10	11	12
A	0.01	0.01	0.01	0.01	0.01	0.01	0.01	0.02	0.02	0.03	0.03	0.04
B	0.01	0.01	0.01	0.01	0.01	0.01	0.01	0.02	0.02	0.03	0.04	0.05
C	0.01	0.01	0.01	0.01	0.01	0.01	0.01	0.02	0.02	0.03	0.05	0.06
D	0.01	0.01	0.01	0.01	0.01	0.01	0.01	0.02	0.02	0.03	0.14	0.33
E	0.01	0.01	0.01	0.01	0.01	0.01	0.01	0.02	0.02	0.03	0.14	0.33
F	0.01	0.01	0.01	0.01	0.01	0.01	0.01	0.02	0.02	0.03	0.05	0.06
G	0.01	0.01	0.01	0.01	0.01	0.01	0.01	0.02	0.02	0.03	0.04	0.05
H	0.01	0.01	0.01	0.01	0.01	0.01	0.01	0.02	0.02	0.03	0.03	0.04

Direction of attack ⟶

Figure 12.3: xThreat Pitch Map

Consider the above graphic which divides the pitch into 96 squares: 12 long and 8 wide. Each square has been attributed with an Expected Threat (or xT) value which represents the percentage chance of a goal being scored from the possession when the ball is in that location, based on historical data. For example, having the ball just outside your own penalty area means you have less than a one per cent chance of scoring from that move, while having the ball in your opponent's six-yard box means that figure rises to 33 per cent. Expected Threat simply offers a numerical representation of the danger levels we intuitively feel as football fans as the ball moves around the pitch. This grid forms the basis of xT and can be used in a number of ways.

The main information which can be harvested from xT is the way players increase their team's chances of scoring via passing, dribbling, or crossing the ball into certain areas of the pitch. A player making a pass from F4 (0.01) into G10 (0.03) has increased his team's chances of scoring by 2 per cent - or 0.02(xT). A player successfully dribbling the ball from C10 (0.03) to D11 (0.14) has increased his team's chances of scoring by 0.11(xT). Expected Threat doesn't require a goal at the end of it like assists do, or even a shot at the end of it like Expected Assists do. The value of each pass can be taken in isolation based on its own merit. Any pass or dribble can be assessed purely on where it started and where it ended. You can even assign value to players receiving a pass in a certain area – passing is a two-way street and requires a player to be in space in a good position just as much as it requires vision from the passer. Some models will attribute a player receiving a pass in F11 (0.05) from G9 (0.02) with 0.03(xT) because he played a hand in progressing his team into this more valuable area. Other models will use Expected Threat to penalise defenders who make mistakes. A centre-back who loses the ball in a position worth 0.01(xT) and gifts it to an opponent striker in a position worth 0.28(xT) to the other team will accrue -0.29(xT). It can also reward defensive players for making tackles or interceptions which prevent opposing players from entering dangerous areas of the pitch.

Suppose we want to identify which players are best at progressing their team into dangerous situations. Figure 12.4 shows the leaders in Expected Threat added from passes and dribbles during the 2020/21 Premier League season. Crosses and pass receptions aren't included in this dataset as they skew the rankings towards cross-heavy full-backs and target men. Negative xT actions such as backwards passes and situations where the ball is lost have also been removed. Jack Grealish topped the rankings in what was his final season at Aston Villa before signing for Manchester City for a British record fee of £100 million. His ball-progressing skills meant he added an average of 0.35(xT) per 90 minutes

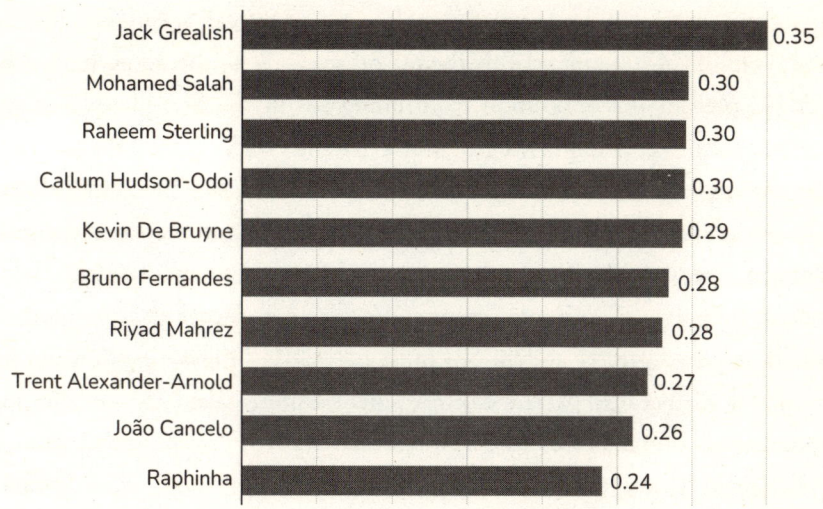

Figure 12.4: Open-play Expected Threat Per 90 Minutes of Action, Premier League 2020/21

he completed. Grealish's dribbling ability and incisive passing around the penalty area certainly passed the eye test as well.

The Expected Threat data in Figure 12.4 can be broken down into its constituent parts: dribbling and passing. Raheem Sterling, then playing for Manchester City, was the most threatening dribbler during 2020/21. His ball carries added 0.18(xT) per 90 minutes of action, pipping second-placed Grealish (0.15). Bruno Fernandes collected the most xT per 90 from passes (0.22), with Trent Alexander-Arnold, João Cancelo and Grealish all adding 0.20(xT). Clearly, Expected Threat gives a neat summary of which players are creating danger and can be put to use by recruitment teams or coaches carrying out opposition scouting.

The problem with Expected Threat is how it perceives actions in the middle of the field. There is a great deal of jeopardy if a defender loses the ball in the defensive third, and a great deal of reward if an attacker makes a brilliant pass in the attacking third. In the middle third, xT demonstrates little regard for either risk or reward. In other words, actions in the middle of the park do little to alter a team's chance of scoring or conceding. This has led some analysts to question the importance of midfielders. Some have argued it validates the spending of more money on attackers and defenders as these players allegedly have more impact on scoring and preventing goals. Others reserve scepticism for this theory – midfield strength is likely the reason a team is able to access high-value areas (or prevent the opposition from doing so) in the first place. And just because an attacker dribbles the ball past a defender high up the field and turns a 10 per cent chance of scoring into a 40 per cent chance of scoring doesn't make the action any more skilful than a midfielder doing so deeper on the pitch turning a 1 per cent chance into a 2 per cent chance. These two take-ons clearly produce different xT outputs, despite the fact the deeper dribble might've actually required greater ability. We must be careful how we portion out the rewards of xT because players can only demonstrate talent in the roles and opportunities afforded to them.

Expected Threat was first conceptualised by Karun Singh, who devised a model which looked at how the location of the ball on the field altered the probability of a goal in that phase in the manner outlined above. Singh has since been hired by Arsenal, the team he has supported since childhood, but his model has been replicated and advanced by several analytics companies, football clubs and solo analysts. The umbrella term for this type of methodology has become 'Expected Possession Value' (xPV) and the more advanced of these models now account for factors such as the speed of the attack, the events leading up to the possession, whether the action is a through ball, dribble, cross, tackle, and so on. Gone are the days where the flashlight of football analytics shone solely on those players who scored goals and took shots. Now every action on the football pitch is being logged, scrutinised, and assigned a value.

JUSTICE TABLES

Arsenal managers have been responsible for two of the most influential mentions of xG in a post-match interview. The first was by Arsène Wenger in November 2017 after his side had been defeated 3-1 by Manchester City. The Frenchman became the first manager to publicly cite the Expected Goals scoreline of a game to defend the performance of his players. Manchester City had accumulated 0.70(xG) to Arsenal's 0.60(xG) and the final score had been unreflective of the true balance of the match. The reaction to his comments was less than favourable.

The British media was hard-hitting in its criticism of Wenger and, in particular, of xG. Several key figures seemed to mistakenly confuse xG for a metric which tells you how many goals each team was expected to score *before* the game took place. They branded xG as 'useless' and mocked the stat. Despite the backlash, the ordeal certainly raised awareness of xG and brought the metric to the forefront of mainstream football discourse.

Fast-forward three years to December 2020. Expected Goals had become a widely accepted metric and was gaining popularity throughout the wider football community, but another managerial interview, this time conducted by Mikel Arteta, was about to spark another bout of confusion and criticism. A 2-1 loss to Everton was Arsenal's seventh consecutive Premier League game without a win and Arteta was feeling the pressure. 'Last year we won against Everton with a 25 per cent chance of winning, we won three-two,' the Arsenal boss stated. 'Last weekend against Everton, it was a 67 per cent chance of winning, any game in Premier League history, and a nine per cent chance of losing, and we lost. Three per cent against Burnley, and we lost. Seven per cent against Spurs, and we lost.'

The interview was a mess. Pundits and fans were left confused as the Arsenal boss listed off a series of seemingly random probabilities of his team's chance of winning their last few matches. Arteta did a poor job of explaining what on earth he was on about, but the thinking behind the statements was actually grounded in cutting-edge analytics: using xG data to generate post-game win expectancy and, in turn, using those numbers to separate the team's performance from the result, just like Arsène Wenger had done three years earlier. Let's take a deeper dive into Arteta's comments.

Arsenal's winless run of results is shown in Figure 12.5, alongside the xG scoreline from each game. The Expected Goals figures in brackets tell us how many goals we could have expected each team to score given the quality of shots they generated. But we can go deeper. Another use of xG is to simulate who was likely to win a game, based on the quality of chances created. Pretend each shot is actually the flip of a coin, but the coin is weighted depending on how good the chance is. For example, a shot worth 0.50(xG) would represent a fair coin that lands on heads half of the time. Alternatively, a penalty is

Opponent	Result	xG Score	Arsenal Win %	Draw %	Opponents Win %	Expected Points
Aston Villa	L	Arsenal (1.57) 0-3 (2.40) Aston Villa	21%	25%	54%	0.88
Leeds	D	Leeds (2.19) 0-0 (0.74) Arsenal	12%	20%	68%	0.56
Wolves	L	Arsenal (0.83) 1-2 (2.33) Wolves	12%	24%	64%	0.60
Tottenham	L	Tottenham (0.27) 2-0 (0.61) Arsenal	52%	36%	12%	1.92
Burnley	L	Arsenal (2.06) 0-1 (1.30) Burnley	64%	21%	15%	2.13
Southampton	D	Arsenal (0.65) 1-1 (1.12) Southampton	19%	33%	48%	0.90
Everton	L	Everton (0.74) 2-1 (1.53) Arsenal	57%	29%	14%	2.00
Points:	2				Total Expected Points:	8.99

Figure 12.5: Arsenal's Form Before Arteta's Controversial Interview, Premier League 2020/21

traditionally converted on 77 per cent of occasions, so the 'penalty coin' would land on heads roughly three in every four flips. The coin of a shot worth 0.01(xG) would land on heads just once every 100 flips. Now suppose that every time a coin lands on heads, the team flipping it scores a goal. By 'flipping' all of the 'coins' in a match, we can simulate how many goals a team would score on average.

Take Tottenham's chances against Arsenal in the 2-0 win outlined in the above graphic. Spurs' shots had xG values of 10, 7, 4, 3, 2 and 1 per cent respectively, giving Tottenham an overall xG total for the match of 0.27(xG). Arsenal, on the other hand, created opportunities worth 0.61(xG). The match plays out. Arteta flips his coins, Mourinho flips his coins and the winner of the contest is the coin-flipper who has heads come up the most often. By repeating this exercise thousands of times, we can 'simulate' the match and work out the share of coin-flipping contests that Arteta wins.

For example, we might play out the coin-flipping contest between Everton and Arsenal one thousand times and find that Arsenal win on 570 occasions. Thus, we can say Arsenal's post-match win expectancy was 57 per cent. The model also tells us a draw would happen 29 per cent of the time and Everton would win 14 per cent of the simulations.

Once you have the probabilities of each outcome occurring, you can use our old friend Expected Value to work out the number of points each team could have expected to take from the match. For example, the following equation calculates Arsenal's Expected Points from the Everton clash:

Expected Points = (Points for Winning x Chance of Winning) + (Points for Drawing x Chance of Drawing) + (Points for Losing x Chance of Losing)

Expected Points = (3 x 57%) + (1 x 29%) + (0 x 14%)

Expected Points = 1.71 + 0.29 + 0.00

Expected Points = 2.00

So, if this match was played hundreds of thousands of times in the exact same way, Arsenal would collect an average of two points per game. You can start to see what Mikel Arteta was getting at with his comments. He claimed that against Everton, 'It was a 67 per cent chance of winning, any game in Premier League history, and a nine per cent chance of losing, and we lost.' Note that Arsenal's xG model differs from the one used to produce the figures above, which is why the figures he quoted are slightly different from the ones shown in Figure 12.5 The club's internal model clearly rated Arsenal's xG performance more highly than our model does, as we give the Gunners a 57 per cent post-game win expectancy. Regardless, the methodology for turning xG into Expected Points remains the same in both cases.

The data shows that Mikel Arteta had cause to be frustrated. Arsenal picked up two points over the course of the seven matches, but their performances merited them 8.99 Expected Points. In other words, they 'deserved' to collect seven more points over this series of games than they did in actuality. The latter half of this run, in particular, saw them perform well – they were more likely than not to defeat Tottenham, Burnley and Everton given the chances they created and conceded in these matches. In late 2020, the dice of fate were coming up snake eyes for Arteta and his team.

But the thing about bad luck is that it runs out eventually. Figure 12.6 shows the seven Arsenal results immediately following Mikel Arteta's controversial interview in December 2020. The club's form turned on its head and they went on a seven-match unbeaten run, collecting 17 points from a

Opponent	Result	xG Score	Arsenal Win %	Draw %	Opponents Win %	Expected Points
Chelsea	W	Arsenal (2.09) 3-1 (2.42) Chelsea	28%	26%	46%	1.10
Brighton	W	Brighton (0.89) 0-1 (1.36) Arsenal	51%	30%	19%	1.83
West Brom	W	West Brom (0.85) 0-4 (4.15) Arsenal	96%	3%	1%	2.91
Crystal Palace	D	Arsenal (0.62) 0-0 (0.77) Crystal Palace	24%	44%	32%	1.16
Newcastle	W	Arsenal (2.58) 3-0 (0.16) Newcastle	92%	7%	1%	2.83
Southampton	W	Southampton (1.30) 1-3 (1.84) Arsenal	79%	15%	6%	2.52
Man Utd	D	Arsenal (0.97) 0-0 (1.87) Man Utd	14%	24%	62%	0.66
Points:	17				Total Expected Points:	13.01

Figure 12.6: Arsenal's Form After Arteta's Controversial Interview, Premier League 2020/21

possible 21. The team's performances improved, as they collected 13.01 Expected Points compared to the 8.99 they'd accrued over the previous seven games. Note the particularly dominant performance against West Bromwich Albion whereby they would only lose the game 1 per cent of the time. This is the exact same chance Brighton had of losing in the matches against the same opponents and Crystal Palace that were referenced in the opening paragraphs of Chapter 1. The Seagulls managed to win neither of these games, showing the extent of their xG misfortunes in 2020/21.

As well as their performances improving, Arsenal's luck also swung on its head as they outperformed expectation by 4.01 points over this run. Maybe their strikers hit a purple patch of finishing, or perhaps the ball fell slightly less invitingly for opposition players to strike at goal. Whatever the case, the xG gods had answered Arteta's plea for a change of fortune. Unsurprisingly, the Arsenal boss didn't quote the post-match win expectancy figures after his side's undeserved victory over Chelsea or lucky draw against Manchester United. Managers usually reserve xG references for when bad luck befalls their team, not when fortune swings their way.

The Expected Points for each team can be added up to form an alternative representation of the league table. These Expected Points tables have come to be known as 'Justice Tables' because they give a fairer view of where each team deserves to rank based on the quality of their performances. Because Justice

Tables are based on xG rather than goals, they strip out the damaging effects of chance and randomness. Figure 12.7 shows an example of a Justice Table comparing the actual and deserved finishing position of the Premier League teams at the end of the 2022/23 season. West Ham, Southampton and Leicester can be counted among the unluckiest sides, while Fulham and Arsenal collected more points than their performances merited.

Expected Points	Expected	Actual
89		
88		MCI
87		MCI
86		
85		
84		ARS
83	MCI	
82		
81		
80		
79		
78		
77		
76		
75		MUN
74		
73	ARS	
72	NEW	
71	BRI	NEW
70		
69		
68		
67	LIV	LIV
66	MUN	
65		
64		
63		
62		BRI
61		AVL
60		TOT
59		BRE
58	BRE TOT	
57		
56		
55		
54	AVL	
53		
52	WHU CHE	FUL
51		
50		
49		
48		
47		
46		
45	LEI	CRY
44	CRY	CHE
43		
42		
41	EVE LEE	WOL
40		WHU
39	FUL	BOU
38		NOT
37		
36	SOU	EVE
35		
34	NOT WOL BOU	LEI
33		
32		
31		LEE
30		
29		
28		
27		
26		SOU
25		

MCI	Manchester City
ARS	Arsenal
NEW	Newcastle United
BRI	Brighton & Hove Albion
LIV	Liverpool
MUN	Manchester United
BRE	Brentford
TOT	Tottenham Hotspur
AVL	Aston Villa
WHU	West Ham United
CHE	Chelsea
LEI	Leicester City
CRY	Crystal Palace
EVE	Everton
LEE	Leeds United
FUL	Fulham
SOU	Southampton
NOT	Nottingham Forest
WOL	Wolverhampton Wanderers
BOU	Bournemouth

Figure 12.7: Premier League Justice Table 2022/23

A common misconception with Justice Tables is that the Expected Points used to form them work in the same way as points do in the actual league table. For instance, people think a dominant xG win grants you exactly three points, a close xG scoreline gives both teams one point and a heavy xG defeat means you accrue zero points. These people often ask what the benchmark is for these outcomes, but this isn't how it works. Justice Tables are generated by adding up the Expected Points totals in the manner outlined above. A team who dominates their opposition might earn around 2.70(xP), rather than exactly 3(xP). If they have a tight game the next week and earn 1.10(xP), then their total Expected Points from the two matches will be 3.80, and so on and so forth.

Sarah Rudd works for StatDNA, an analytics company purchased by Arsenal back in 2012. StatDNA serve as a sort of internal data consultancy and advise the club on a number of strategic measures, from recruitment to opposition scouting. One of Rudd's responsibilities is looking after Arsenal's post-game win expectancy model, the one that formed the basis of Mikel Arteta's notorious press conference. Rudd has since lifted the lid on the club's use of the algorithm. 'What we want to do in the post-match report is separate the result from the process and try to give the coaching staff insight into how well we played,' she said. 'Were we reaching our objectives but then lost because we got unlucky?' Rudd's comments neatly encapsulate the value of viewing football results through the lens of advanced metrics. It's a lens an increasing number of football clubs are using to view the game through.

As expected, Brentford arrived early to the Expected Points party. Their manager, Thomas Frank, once told the press, 'I don't talk about wins and losses. I talk about performances.' The subtext of this comment reads, 'I don't care about points, I care about Expected Points.' Lee Dykes, Brentford's Technical Director, has backed up his manager's statements by saying, 'It's about getting better every year. Not by league position, but by the measurements we take.' Subtext: 'We don't care about the league table, we care about the Justice Table.' Performance (or 'process,' as Arsenal have taken to calling it) is what matters in football. This is what you control. The performance of your team is muddled in with a great deal of luck – the product of this combination is the final result. If Brentford face a difficult opponent and lose the game 2-0 but win on xG, their manager will be told that the team have been upgraded in Smartodds' mathematical rating of the club and that he shouldn't worry. Keep performing well and the wins will come.

Phil Giles gives an example of such an occasion. Brentford's Director of Football remembers, 'In our first season in the Premier League, we lost 1-0 to Chelsea but we absolutely battered them.' They did: the xG scoreline read **Brentford (1.88) 0-1 (0.28) Chelsea**. The Bees hit the woodwork on a couple of occasions and Edouard Mendy, Chelsea's goalkeeper, was named Player of the

The Expected Goals method argues against the old adage, 'All's well that ends well.'

Match. Brentford had a post-match win expectancy of 80 per cent and collected 2.59 Expected Points from the game. '[Brentford owner] Matthew Benham turned to me and said, 'I think that's our best ever performance. The best we've ever played.' He was really happy with it,' noted Giles. The flip side of this, of course, is that when Brentford win undeservedly the celebrations are sometimes cut short by a text message from the owner invoking disappointment at the performance. The Expected Goals method argues against the old adage, 'All's well that ends well.' You might win a game or series of games, but all might not be well with your team and their performances.

Brentford aren't the only ones to harness the power of Expected Points. We've seen that Liverpool's research team used this type of model to demonstrate how unlucky Jürgen Klopp had been at Dortmund, and subsequently hired him as their manager. Brighton's success has been founded on the similar sort of Justice Tables and global ranking systems as Brentford, while other teams across Europe will no doubt have their own Expected Points systems. And of course it's not just football clubs which benefit from the perspective given by Expected Points. Professional bettors use this system to guide their betting and identify teams who might be undervalued by the bookmakers.

Post-match win expectancies can be used to analyse football results in another way. Figure 12.8 compares the post-match expectancy to the implied pre-match chance the bookmakers gave Arsenal of winning the seven games prior to and after Mikel Arteta's press comments. The bookies' odds represent the best forecasts of match outcomes given they're the product of hundreds of thousands of individual predictions, including those by the industry-leading betting

Opponent	Result	Pre-Match Win %	Post-Match Win %	Differential
Aston Villa	L	59%	21%	-38%
Leeds	D	43%	12%	-31%
Wolves	L	47%	12%	-35%
Tottenham	L	24%	52%	28%
Burnley	L	63%	64%	1%
Southampton	D	42%	19%	-23%
Everton	L	34%	57%	23%
Chelsea	W	23%	28%	5%
Brighton	W	43%	51%	8%
West Brom	W	62%	96%	34%
Crystal Palace	D	63%	24%	-39%
Newcastle	W	69%	92%	23%
Southampton	W	47%	79%	32%
Man Utd	D	34%	14%	-20%
Average:		47%	44%	**-3%**

Figure 12.8: Arsenal's Pre- and Post-Match Win Expectancies, Premier League 2020/21

consultancies who have mastered the prognostication of football matches. The first thing this graphic tells us is that Arsenal's fixtures did get slightly easier in the seven matches following Arteta's comments (49 per cent average win probability) than they were in the seven preceding matches (45 per cent). This goes some way to explaining the improvement in their Expected Points output over this period.

We can also compare the pre-match and post-match probability of winning to better understand how well the team played. Take the Everton game which preceded Arteta's infamous comments. The betting markets were giving Arsenal a 34 per cent chance of winning this game before it kicked off. Arsenal fell to a 2-1 defeat, but their post-game win expectancy was 57 per cent. Thus, we can say the Gunners' performance over the course of the match increased their chance of victory by 23 percentage points versus expectation. They outperformed what was anticipated of them. Meanwhile, the final game of their seven-match unbeaten run saw them manage a draw against Manchester United, but they played worse than the market thought they would. The average implied bookmaker odds assigned to Arsenal over this entire 14-game period was 47 per cent, but they only managed to achieve an average post-game win expectancy of 44 per cent.

Thinking probabilistically is crucial when assessing football performance. Things aren't always black and white; they're often a particular shade of grey. We need to learn to think in probabilities: supplementing goals with Expected Goals, points with Expected Points, and wins with post-match win expectancy. This allows us to develop a deeper, clearer perspective of the ability of teams. Consider an xG parallel universe, where each side's success and failure rests solely on their Expected Goals performance. Here are some of the key headlines from this alternate reality:

- Manchester City won every Premier League title between 2016/17 and 2022/23. In each of these seven seasons, Pep Guardiola's team amassed the most Expected Points in the league and topped the Justice Table. In our own universe, this streak has been broken up by Chelsea's victory in 2016/17 (when the Blues overperformed by 17 points and City underperformed by 7 points), and by Liverpool's victory in 2019/20 (when the Reds overperformed by 24 points and City underperformed by 6 points).
- The 2015/16 Premier League title was won by Arsenal, not Leicester. Riyad Mahrez scored 12 goals instead of 17, while Aaron Ramsey and Theo Walcott scored nine and eight goals respectively instead of five each. If these players, as well as the rest of the players in each team's squad, converted chances at their expected rate then Arsenal would have won the league by seven points.
- Brighton & Hove Albion claimed a Champions League spot in 2022/23 instead of Manchester United. The Seagulls produced the most remarkable underdog story in recent Premier League history, placing fourth despite possessing the third-smallest wage bill (remember that Leicester's title win didn't happen in this xG universe).
- Jürgen Klopp's Dortmund weren't bottom of the Bundesliga entering December 2014. Instead, they were fourth. The German manager might not have left the club at the opportune moment for Liverpool to hire him.

- Every Serie A winner between 2017/18 and 2021/22 would have been different. In our universe, Juventus won the league three years straight from 2017/18 to 2019/20, before Inter Milan and AC Milan triumphed the following two seasons. However, in the xG universe the league was won by Napoli in 2017/18 and 2018/19, before Atalanta arrived on the scene and lifted the Scudetto in 2019/20. In our reality, 2020/21 was the first season Juventus didn't win Serie A since 2010/11, but this was also the first time in four years that they actually deserved to do so. They finished fourth in our universe, but narrowly pipped Inter Milan and Atalanta on Expected Points. The real Serie A was won by AC Milan the following season, in 2021/22, but the xG Serie A was actually claimed by their close rivals Inter Milan.

- La Liga's recent seasons in the xG universe have been dominated by Barcelona, who won six titles in the eight years between 2014/15 and 2021/22. Real Madrid managed the other two, while Atlético Madrid's sole title-winning season in our universe in 2020/21 actually saw them come fifth in the world of xG, scoring nine fewer goals, conceding 10 more, and collecting 20 fewer points.

- Over in Ligue 1, Lyon can be disappointed not to have claimed their first league title in 13 seasons in 2020/21. Their performances merited enough points to narrowly beat Paris Saint-Germain, Monaco, and actual champions Lille.

The xG parallel universe represents what would have taken place if every game from our own universe was simulated thousands of times as if each shot was an xG coin flip. Note that this alternate reality doesn't account for the impact of game state. How much you believe a team 'deserved' to win a game based on xG and xP will be determined by how distorting you believe the effects of game state to be. A team who goes 2-0 up early in a match might look to sit deep and retain their lead, meaning their xG might not be as generous as if they'd taken the lead later in the game. Thus, it could be argued that teams who take early leads and then seek to contain the opposition might be unfairly penalised in any Expected Goals or Expected Points calculations. While the game state argument does carry some credence, it should also be remembered that xG should theoretically be easier to accumulate when a team is in a winning position because bigger chances can be created when the opposition throws players forward. Manchester City certainly didn't collect the most Expected Points seven seasons in a row between 2016/17 and 2022/23 by sitting back and defending when taking the lead. Good teams should be able to outperform their opponents no matter the game state.

Justice Tables make it easy to compare the performance of teams in the same league, but can we go one step further and compare the quality of clubs from different corners of the world? In other words, is there a way we can rank all professional football teams as if they played in one giant league table? In short, there is. Suppose West Ham are playing Villarreal and you wanted to know the chance of victory for either side. These two teams have never played a competitive game in their history, so there's no prior form to go off. What we do know is how good West Ham are compared to top Premier League teams like Manchester City, and we know how good Villarreal are relative to top La Liga teams like Barcelona. If we can map these relationships then we can gain a strong understanding of the relative quality of West Ham and Villarreal.

A second way of comparing teams from different leagues is to build a Player Contribution Model. Such a model gives every player in the world a rating, often split between an attacking rating and a defensive rating. Smartodds rate players using a star system – a player of vital importance to a team would get five stars, a player who ranked poorly on the company's KPIs would get one star. These models often rate players according to the strength of the club they play for, how they are utilised, the number of minutes they play, the number of goals their team scores and concedes when they're on the pitch, as well as more advanced metrics like Expected Goals and Expected Assists. Ultimately, a team is a collection of 11 players. An analyst who knows the ability of the 11 players a team fields in any given game can accurately assess their chance of victory. If Kylian Mbappé gets injured warming up pre-game, you'll see the bookmakers' odds drift for his team. Professional bettors aim to be the first ones to discover team news or any alterations to the starting line-up in an attempt to find an edge. The downside to Player Contribution Models is they miss the role of team cohesion or tactical set-up. These aspects can make a team better than the sum of its parts.

Brentford and Brighton's data consultancies use a blend of these two methods – cross-division mapping and Player Contribution models – to calculate the quality of each and every professional team. European football is like an enormous spider web, and each match which takes place between two clubs is a single thread. By plugging the information from each thread into their models, analysts can build one big league table comprising every team throughout the world. As it so happens, West Ham and Villarreal sit next to each other in our own version of the Global Justice Table at the time of writing, harbouring 43rd and 44th position respectively. Whenever your club plays a match, you naturally consider the implications on your league standing. 'If we win, and that other team lose, we could move up to seventh place,' the anonymous fan might state. However, within Smartodds' framework, each club is not just competing with teams within its actual division – be it the Premier League, the Championship, Moldova's second division or so on – it is also competing with every other professional football team in the world. A fan with access to Smartodds' model could support Gillingham and excitedly exclaim, 'If we win today, and Ordabasy of the Kazakhstan Premier League lose, we could move up to 675th in the Global Justice Table!'

Actually this is not what the fan would be saying at all. Wins, draws and losses don't count for anything in Benham's system. What counts is the xG you create and concede in each match and, in turn, how this affects the model's rating of your team. The fan would be more correct in exclaiming that, 'If we create a large number of high-quality chances, while simultaneously conceding very few, we will amass a dominant Expected Goals scoreline and thus move higher up in both our domestic xG Justice Table and the global xG Justice Table!'

OTHER ADVANCED METRICS

Football analytics can be likened to the dashboard of a car. You wouldn't drive a vehicle without the various indicator panels situated on the dashboard, just like you wouldn't run a football club without an analytical department. Expected

Goals is the speedometer – this stat tells you how fast you're going – but there are a handful of other useful metrics, proxies, and gauges that light up this dashboard.

Pressing has become an important part of the modern game. The 2010s saw a philosophy shift away from the pass-crazed teams of Pep Guardiola's Barcelona and towards the press-crazed teams of Jürgen Klopp's Liverpool and, to a certain extent, Pep Guardiola's Manchester City. Perhaps the best proxy football analytics that has developed for pressing intensity is known as 'Passes Per Defensive Action' (or PPDA). The metric counts how many passes a team allows an opponent to make before trying to win the ball back via a defensive action such as a tackle, interception, foul, failed foul, or clearance. Importantly, this metric only applies to the attacking 60 per cent of the field (an area roughly aligned with the edge of the semi-circle in your own half).

Let's take Marcelo Bielsa's Leeds United as an example. This team was known for its intense pressing nature and constantly ranked highly for PPDA. In 2020/21 they obtained an average PPDA of 9.3 – the lowest in the Premier League. In other words, they only allowed the opponents to make nine passes before they attempted some sort of defensive action. On the other end of the spectrum that season was Newcastle, who preferred to drop deep into a defensive shape and rarely pressed the opposition high up the field. The Magpies allowed an average of 18.3 PPDA per game over the 2020/21 campaign.

PPDA is a useful proxy, but has two major shortcomings. First, it measures the pressing *intent* of a team, but not their *success*. We know Leeds attempted a high number of defensive actions in a quick period of time high up the pitch, which gives us an insight into their style of play, but doesn't tell us how well they did it. Teams like Liverpool and Manchester City ranked slightly lower (with 10.4 and 11.5 PPDA respectively), but probably pressed better in unison and were more efficient at winning the ball higher up the pitch. The second shortcoming is that it only measures teams – PPDA can't be used to measure the pressing ability of individual players.

'Field Tilt' is another indicator on the car dashboard; a metric which illuminates the territorial dominance between two teams during a game. Field Tilt is to possession what xG is to shooting data; it gives a more powerful view of a fairly surface-level stat. Field Tilt measures the share of possession a team has in the game but only measures touches or passes in the attacking third of the pitch. If there are 100 final-third actions in a game and Team A makes 80 of them, then Team A has a Field Tilt of 80 per cent. The teams at the top of the Field Tilt rankings are the usual suspects: Manchester City, Liverpool and Chelsea inhabited the top three positions in the Premier League in 2021/22 and achieved the same ranking in the Field Tilt table. City averaged 70.3 per cent Field Tilt, while Liverpool and Chelsea averaged 67.4 per cent and 59.9 per cent respectively. Traditional possession statistics tell you who is better at keeping the ball, but not a great deal about the areas they're getting into. Field Tilt measures who is controlling the ball in the area that matters: the final third. As well as comparing teams against one another, it can also be used to identify changes in the way single teams are playing. Newcastle ranked as the fifth-lowest team for Field Tilt in 2021/22 (40.5 per cent) but then improved to the fifth-best team the following season (59.8 per cent). Analysts can dig deeper and identify any changes to the tactics, formation or starting line-up which might be driving their improved performance.

Recent years have seen a useful but slightly less-well-explained metric come to the fore: 'Match Momentum.' Otherwise known as 'Goal Threat,' 'Attack Momentum' or some combination of similar words, these graphics outline the flow of the game. Which team is dominating at certain points and who is causing more threat? Most models do this by measuring the likelihood of a team scoring within the next 10 seconds or so. These are, in essence, slightly more advanced versions of Karun Singh's Expected Threat model. Analysts can calculate the danger levels of certain situations by aggregating individual actions – carries of the ball into the attacking third, passes into the box, and so on – to the team level, allowing them to work out who is creating the most dangerous chances without the need for a shot even taking place. Match Momentum looks at the peak moments of danger for either side and works out the 'attacking threat' for each minute.

Figure 12.9 depicts the Match Momentum graphic which was shown on BT Sport's coverage of Manchester United versus Atalanta in the Champions League Group Stage in 2021. This was a historic moment for this type of analysis because it was one of the first times a television audience were given a glimpse into the flow of the game from an attacking threat perspective. Atalanta carried a great deal of danger during the first half and managed to go in at the break with a two-goal advantage. United were much better after half time, not allowing Atalanta many opportunities to get forward and managing to win the game with three second-half goals. Graphics like this provide an easily digestible snapshot of how much pressure each team applied throughout the match, and can supplement xG and other metrics to provide a clearer image of how a game unfolded.

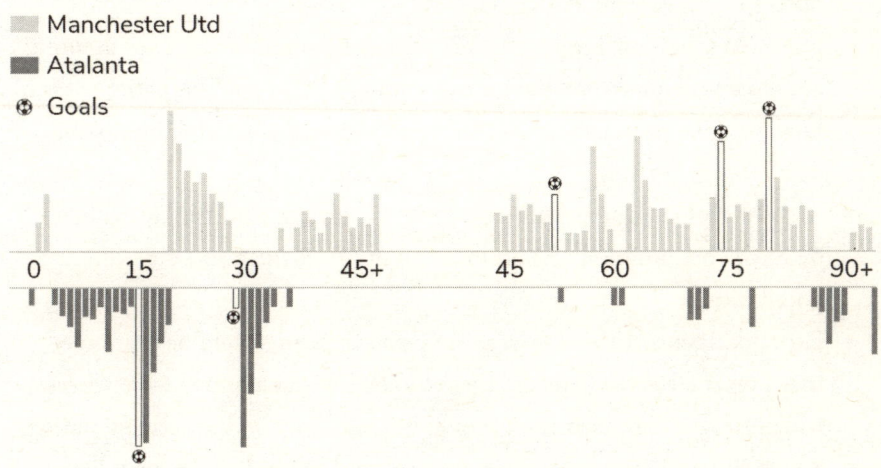

Figure 12.9: Match Momentum, Manchester United 3-2 Atalanta (20 October 2021)

Advanced football metrics are slowly starting to seep into the public consciousness. There's a natural delay between the arrival of a new metric on the analytics scene and the integration of that metric into the media. The tools outlined in this chapter are starting to get the airtime they merit. Stat Perform's Match Momentum graphics have been used in broadcasts by French broadcaster Canal+ during Champions League matches, while Premier League Productions have framed the same metric as 'Goal Threat' and shown in-game graphics of the percentage chance of each side scoring within the last 10 minutes based on their attacking threat. PPDA and Field Tilt have also been used in television broadcasts, particularly by Sky Sports' flagship analysis show *Monday Night Football*. The list of stats shown at full time a couple of years ago pales in comparison to the numerous data points the pundits have at their fingertips in the modern era. Each of these metrics serve to illuminate the sport of football in their own way. From xGOT to Expected Points, from PPDA to Filed Tilt; the dashboard of football analytics is becoming increasingly sophisticated.

CHAPTER SUMMARY

- A whole suite of advanced metrics, such as Field Tilt, have been developed to complement xG analysis.
- Expected Goals on Target (xGOT) tells us how well strikers are shooting and how well goalkeepers are shot-stopping. Expected Threat (xT) tells us how well players are affecting their team's chance of scoring by passing, dribbling or crossing.
- Metrics such as Match Momentum are becoming increasingly utilised by media organisations, resulting in an uplift in their understanding and popularity.
- Expected Points show us how many points a team should have amassed based on their performances. These can be translated into Justice Tables, which show where each team should be positioned in the league standings.
- Smart clubs are using the above metrics, and more, to help them better understand performance and win more football matches.

13

THE DRESSING ROOM

ANALYSING THE HARMONY OF SQUADS

*'Not everything that can be counted counts and
not everything that counts can be counted'*

Albert Einstein

Football's popularity is, in part, a product of its unrivalled ability to generate fresh and intriguing narratives. The chief protagonists in this never-ending soap opera are undoubtedly the players, the global superstars who attract mass interest from all corners of the globe. Lionel Messi and Cristiano Ronaldo have amassed a combined total of over 500 million social media followers, each possessing an audience which dwarfs that of any club. Fans follow their every move both on and off the pitch, from their training regimes to their personal lives. Transfer speculation surrounding the sport's biggest names is unrelenting. The media spotlight which used to shine most brightly on clubs is now fixated on individuals, as seen in the way the game is marketed and the coverage it receives in the media. In a sport becoming increasingly obsessed with players, are we losing touch of what it means to be a team?

According to the media, Argentina didn't win the 2022 World Cup because of their tactical set-up, their team cohesion or any other quality attributable to the collective group. They won it because of Messi. If France had triumphed in the final, Kylian Mbappé undoubtedly would have been hailed as the deliverer of the trophy. This isn't to say that these players didn't play a huge role in the success of their teams, but rather that we overstate the importance of individuals and understate the significance of team performance in our post-match analyses. This applies to players and clubs of all sizes. Primary goalscorers or particularly influential midfielders are often heralded as the key drivers of success, when in reality the credit should be spread more evenly throughout the team.

'Player of the Match' awards are generally given to the player who scores the most goals in a game, but has this player really had the most impact on the match or do they simply benefit from being at the business end of the pitch? A midfielder who works tirelessly off the ball for 90 minutes is arguably more deserving of POM than a striker who does nothing but take one kick which finds the net. The problem is it's difficult to gauge how the output of a single player benefits the collective other than by counting surface-level stats like goals and assists. The same goes with how individual awards such as the Ballon d'Or are handed out. Jorginho placed third in 2021 following Chelsea's Champions League victory and Italy's Euro 2020 triumph, but few would argue Jorginho was anywhere near the top 20, let alone the top three footballers in the world that season. The Italian profited from our inability to separate individual output from team success.

Credit for success is rarely attributed to the set-up and composition of a team: the relationships and interactions within it, the delicate blend of skills and capabilities; the way the recruitment team have brought in certain players to play certain roles; the way the management staff have set the team up tactically and trained the players to perform particular tasks. These more nuanced drivers of performance have a material impact on success, but are rarely mentioned in post-match analysis. Why is this? Well, from an analytical standpoint, these characteristics are much harder to measure. We can easily count the number of goals Messi scores, or the Expected Assists output he provides. It's much harder to model how the interpersonal relationship between Alexis Mac Allister and Ángel di María helped the former to deliver an inch-perfect cross to the latter for the second goal in the World Cup final, or the positive impact the inclusion of a particular squad member might have had on the morale of the dressing room.

Swarm harmonisers provide better opportunities for teammates to showcase their talents.

Some people have an ineffable ability to improve the collective output of everyone around them. These people are known as 'swarm harmonisers' – a phrase which originated from biological studies of the coordinated movement of groups of animals, but which sport has adopted to describe players who enhance their team not only through their own skill, but by improving the ability of those around them. Swarm harmonisers provide better opportunities for teammates to showcase their talents and help a team become better than the sum of its parts without necessarily taking the credit.

Roberto Firmino's statistical output is fairly ordinary. The Brazilian averaged 10 goals and six assists per season during his time at Liverpool despite playing alongside some of the most fearsome wingers in the world. However, the recent development of tracking data has illuminated a previously unrecognised string to Firmino's bow: his ability to create space for teammates by making intelligent runs. Liverpool's analytics department likely discovered this skill of Firmino's long before anyone else. When he was first signed, the recruitment team implored Brendan Rodgers to play him down the middle rather than on the wing because they realised he was more effective and disruptive there. Although Rodgers didn't oblige, Jürgen Klopp's arrival saw Firmino transformed into a 'false nine' – a striker who drops deep, pulls centre-backs out of position, and opens up room for wingers such as Mohamed Salah and Sadio Mané to exploit.

Analysts have developed software which registers where all 22 players are positioned on the field at any point in time. Even when the camera pans away from defenders on the other side of the field, a technique called 'ghosting' can guess where the out-of-shot players are positioned based on historical data provided by full-pitch cameras. Conventional stats like passes, tackles, shots, and interceptions all measure what is happening to the ball. Looking at the game through the lens of these stats is like shining a torch on the action happening where the ball is, while leaving the rest of the pitch in complete darkness. Only a small portion of what happens on the field actually happens near the ball. Tracking data can help us cast a broader light and see what's happening on the rest of the pitch; where players are and where they're moving to. Marrying traditional event data to newly discovered tracking data has proven incredibly effective at revealing the previously hidden talents of players like Firmino. Every time Liverpool make a pass, what is Firmino doing? Researchers found the Brazilian accumulates a high number of 'hidden assists' whereby a goal wouldn't have been scored without his off-ball movement: he did not touch the ball and would not get any credit in the conventional stats but he made a big contribution. Roberto Firmino is a classic swarm harmoniser: his presence makes his teammates better.

N'Golo Kanté is another such example. The two-time Premier League winner played alongside Danny Drinkwater at Leicester and made the Englishman look good enough to secure a transfer to Chelsea. He then played with Jorginho at Stamford Bridge and made him look good enough to come third in the Ballon d'Or. He has played beside Paul Pogba in the French national team and made him look a completely different player to the one who has turned out for Manchester United and Juventus. Wherever Kanté has gone, he's made the midfielders he's played alongside perform to the best of their ability.

Clubs can also use tracking data to assess the 'attitude' of a player: is he willing to run back home when he's out of position or when the opposition is counter-attacking? How fast and how often is he running back to help his teammates defensively? Kanté obviously ranks highly in these qualities, but clubs find it useful to carry out the same assessments for wingers who might not necessarily be required to help out defensively, but do so anyway because they display swarm harmoniser tendencies. Similarly, tracking data unveils whether attacking players keep making forward runs even when they're not being found by teammates. Do they give up, or persevere?

Some analysts believe body positioning and movements is the next frontier of tracking data. We now have automated offsides which tell us when a player's body part is the wrong side of a defender. Can we use similar technology to analyse how creative midfielders like Kevin De Bruyne kick the ball? Or if there's a certain way a winger dribbles that makes them better at beating players? Can these specific movements be taught and learned?

Another way tracking data could revolutionise the sport is in how the defensive positioning of players is analysed. N'Golo Kanté in his prime was regularly commended for making a lot of interceptions, but are there players so good at positioning themselves between opponents that the ball doesn't even get passed near them? These players would arguably be more effective than Kanté, but wouldn't surface as highly in the rankings of current statistical measures such as interceptions, tackles, and the like. Is there a midfielder as good as Kanté who we can't identify yet? Any club who develops the technology to detect such a player will gain a massive edge. Player tracking is still a nascent, emerging field, but any club, data consultancy or analyst who can master it will gain a big advantage over their rivals.

TEAM CHEMISTRY

Of course, some swarm harmonisers benefit their team in a more psychological manner. The presence of a calming influence or a natural leader empowers those around them. Readers might be surprised to see a book which aims to outline the cold, hard science of football exalting the virtues of 'experience' or 'passion.' These qualities have become go-to buzzwords for pundits who struggle to think of anything better to say than 'the team who lost today didn't show enough heart.' But it's undeniable that swarm harmonisers exist on a psychological basis. Anyone who's had a particularly compassionate or inspiring boss can tell you how much of an impact these qualities have on a wider team.

A similarly difficult-to-measure but equally as important characteristic to note is that of 'team chemistry,' which refers to the level of cohesion, communication, and cooperation among team members. When a team has good chemistry, its members work together effectively and are able to achieve their goals more easily. Consider the role of a manager: to pick the team which has the greatest chance of success; or the job of a head of recruitment, which is to assemble the squad with the best chance of winning matches. These are subtly different tasks to simply picking or signing the best players. The point of selection is to maximise the output of the whole, not necessarily to pick the best individuals. A team filled with highly skilful 'bad apples' won't perform as well as one with slightly less talented players who work well together. When team members feel valued and supported by their teammates, they are more likely to perform well and take risks without fear of failure or judgement. This can lead to greater innovation and creativity, as well as a greater willingness to work together to solve problems.

Rowing teams provide a material example of prioritising the collective over the individual. The final eight rowers selected as crew members are rarely the 'best' eight rowers from the squad in terms of raw speed or power. A race isn't the sum of eight separate rowers all tugging individually at rowing machines in a gym. It's the collective endeavour of one team out in the open water. The job of a rowing coach is to select the fastest boat, not the eight fastest individual crew members. Or, as Johan Cruyff reasoned, 'Choose the best player for each position and you'll end up not with a strong XI, but with XI strong I's.'

Lee Dykes, Brentford's Technical Director, provides a footballing example of how to think of your team make-up. Dykes was once talking to a very good friend who was managing a Championship team. The manager was struggling with a left centre-back whom Dykes knew well. When Dykes told him that he needed to sign a new right-winger, the manager looked at him in bemusement and asked what he meant. Dykes explained that the centre-back's biggest asset was his distribution, his ability to play long switches across the field. If the manager played a right-winger who was good in the air, he'd start to get the best out of the struggling centre-back in possession. That's an example of how you have to think about your formation and squad models. Each footballer isn't an individual asset, but rather one component in a wider machine.

It's important to consider how the skillsets of the players in your squad complement one another. If you have a right-winger who is great going forward but poor defensively, perhaps you need to play a more defensive-minded central midfielder on the right-hand side in order to cover the winger. If you have one centre-back who is great in the air and on the ball but is incredibly slow, you might need to play a quicker central defender alongside him. Each team is a delicate ecosystem filled with various talents. It's a game of yin and yang whereby balancing the strengths and deficiencies of the squad members is crucial. The same is true for coaching staff. One backroom staff member might be a 'laptop coach' who is analytical, introverted and spends a lot of time in his office. Another one might be a 'training-ground coach' who spends a lot of time out on the training pitches and among the players. A good backroom staff will have a blend of all sorts of personalities and skillsets. Of course, there are some core fundamentals which ideally everyone would possess – conscientiousness, open-mindedness, self-awareness, and so on – but the ideal team will be a diverse portfolio of backgrounds, experiences, and talents.

Suppose you're making a watch. There's no point in sourcing the most expensive cogs, springs and mechanisms on the market if they don't all fit together in unison. A watch built from cheaper materials which were specifically obtained to fit together will work better. Football clubs could learn a thing or two from this analogy. Perhaps the biggest recent offenders of trying to fit expensive square pegs into round holes have been Manchester United. Since Sir Alex Ferguson left in 2013, the club have spent well over £1 billion on player transfers but have come nowhere near to competing for the Premier League trophy. United achieved an average league position of lower than fourth place over this period despite possessing the largest revenues in the league. They are the antithesis to clubs such as Liverpool, Brighton, and Brentford.

The responsibility of maintaining Manchester United's position at the pinnacle of English football in the post-Ferguson era fell to Ed Woodward. Having been hired as head of United's commercial operations in 2007, Woodward did a successful job of growing the club's revenues via a number of sponsorship deals. When Ferguson retired, Woodward was promoted to chief executive. His new role basically made him the head of football operations, a job he wasn't particularly qualified for. Woodward's experience prior to arriving at United had been in banking, where he'd mastered the art of making deals and growing revenue lines.

What he didn't have experience in was how to run a football club. A series of high-profile but costly signings hallmarked the Woodward era: purchases such as Romelu Lukaku, Alexis Sánchez, Henrik Mkhitaryan, Ángel Di María, Memphis Depay, and Paul Pogba. The club also brought in a number of older, declining players on huge wages who blocked the development of young talent. The recruitment of Zlatan Ibrahimović (34 years old), Bastian Schweinsteiger (30) and Cristiano Ronaldo (36) typified the lack of long-term plan in the post-Ferguson era. The presence of big egos and an absence of discernible swarm harmonisers meant the dressing-room atmosphere was frequently described as 'toxic'.

The problem wasn't necessarily that Woodward's signings were bad players – many of them achieved success before arriving at United and again after leaving. The problem was that Manchester United had no defined style of play. The players weren't recruited to play a particular role within a particular system, they were signed solely because they were talented footballers. No player demonstrated this point more than Paul Pogba, who played virtually every position in midfield and attack during his spell at Old Trafford. The players weren't signed to play within a pressing system, nor a possession-based system, nor a counter-attacking system. David Moyes, Louis van Gaal, José Mourinho and Ole Gunnar Solskjaer were the first four managers appointed in the years following Ferguson's departure, but there are very few tactical similarities from one to another. Remember how Brighton recruited Roberto De Zerbi because he favoured a similar tactical set-up to the one the club had established under Graham Potter, allowing a consistent transition from one manager to the next. Manchester United jerked from one manager to another without any thought of playing style or consistency of approach. The components of the watch didn't fit together.

To use another analogy, Manchester United used to be the best restaurant in the land, winning multiple awards year upon year. Then the long-standing head chef left and was followed by a number of different successors, each one bringing with him a different recipe book. One set the cooks up to serve Scottish haggis, the next Dutch pancakes, the next Iberian pork, the next Norwegian meatballs. The restaurant bought lots of new ingredients in the process, but this cost them a lot of money and meant they ended up with a stock room full of contrasting flavours. The whole business was a mess and lost all of its acclaim, much to the delight of its many rivals up and down the street.

The hiring of Ralf Rangnick was a recognition the restaurant needed to be more aligned. They needed to settle on a particular cuisine (style of play) and buy the appropriate players (ingredients) to cook better dishes (put in good performances) and end up winning awards (trophies) again. There are signs that happiness could return to the restaurant in the near future. They have always had more resources and deeper pockets than the competition. They have a bigger catchment area and greater footfall. If they can get their ducks in a row, they will surely become one of the most successful restaurants in the land once again.

TURNING INDIVIDUALS INTO A COLLECTIVE

Smart teams will turn the responsibility of the individual into that of the collective. Consider the Oakland Athletics, who turned the individual action of a batter hitting the ball into the collective objective of multiple batters getting on-base. When a player left the team, Billy Beane would speak of the need to 'reproduce them in the average.' In other words, he would find another cog to fit into the well-oiled machine that was his baseball team. Sports teams don't innovate by signing a single specific world-class player, they do so by changing the whole team's approach and acquiring players who play the game in a different, better way to everyone else. Sean Dyche's Burnley, perhaps one of the most successful Premier League teams relative to their budget in recent memory, were also subscribers to this theory: their 4-4-2 formation was king. Players would come and go but the system would stay the same. Meanwhile, Liverpool's famed strategy of pushing the wing-backs high and dropping the striker into midfield was the driving force behind their success under Jürgen Klopp. Most, if not all, prosperous teams have a blueprint of how they play which takes precedence over any individual player.

Certain clubs are making vast strides in how they measure the performance of the entire team on the pitch at any one time. Historically, football data has represented the output of individuals on the field of play. Passes, tackles and shots are all solo actions. Possession has been the primary indicator of a team's ability to keep the ball, but this statistic is shallow and outdated. Enter Tim Waskett, an astrophysicist, and Will Spearman, who has a doctorate in philosophy. These two members of Liverpool's data science team helped develop the notion of 'Pitch Control.' The technical definition of Pitch Control is, at a given particular location on the pitch, the probability that a player could control the ball assuming it were at that location. It's essentially a way of quantifying how teams control space on the pitch.

Figure 13.1 shows a single frame of player-tracking data. The white dots represent players from Team A, who are shooting right to left, while the black dots represent their opponents. The regions of control are shown by the shades of grey spreading across the pitch. Any dark-grey zone is space a Team B player would be able to access first if the ball was passed there, while the opposite is true for the light-grey zones. Around both goalkeepers, their team has complete control. The midfield is an ever-changing situation as players move dynamically among one another.

'Control' of the pitch is defined by how close each player is to the ball, as well as their direction and speed of movement. This information is all gleaned from player-tracking data. Of course, little is certain in sport. Waskett and Spearman

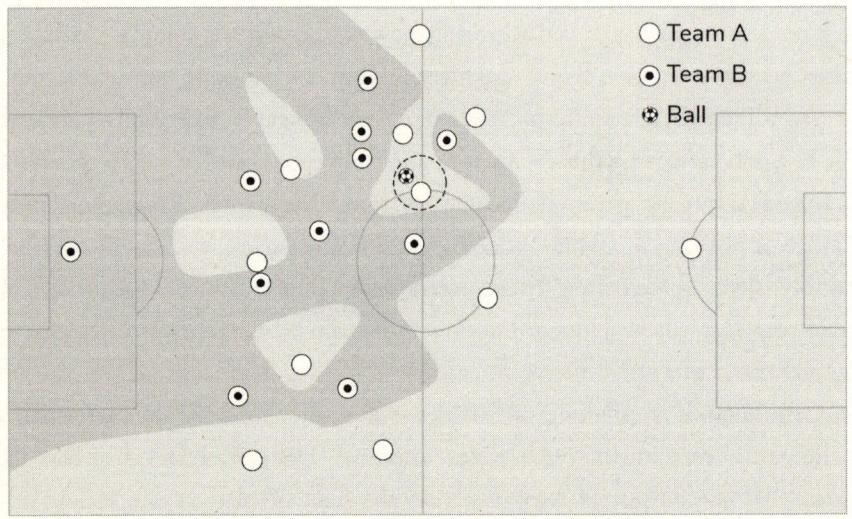

Figure 13.1: Example of a Pitch Control Model

wanted to account for this uncertainty – rather than saying the player with the lowest time to intercept was in control of the ball, they added a model which worked out the probability of each player being able to get to the ball first. The darker the shade, the greater the control of that area by that team.

The snapshot of the game as displayed in Figure 13.1 can be improved upon. Layering in the direction and speed of the ball will give a clearer view of who is more likely to gain control of it. Clearly, the areas close to the ball are more relevant. If the ball is near the halfway line, a goalkeeper having control of almost his entire half isn't as relevant as a midfielder finding 10 yards of space in the middle of the park or a striker getting into a dangerous position between two defenders. Analysts can also layer in the importance of various areas of the field by using a model similar to the Expected Threat models we explored in the previous chapter. Having control by the centre circle isn't as valuable as having control in your opponents' box.

Figure 13.2 shows a different scenario which incorporates the movement of the ball, only shows the relevant areas of the pitch, and layers in the threat levels from various pitch locations. Team B are clearly counter-attacking and the dark areas are the places they can get to quicker than Team A. The player

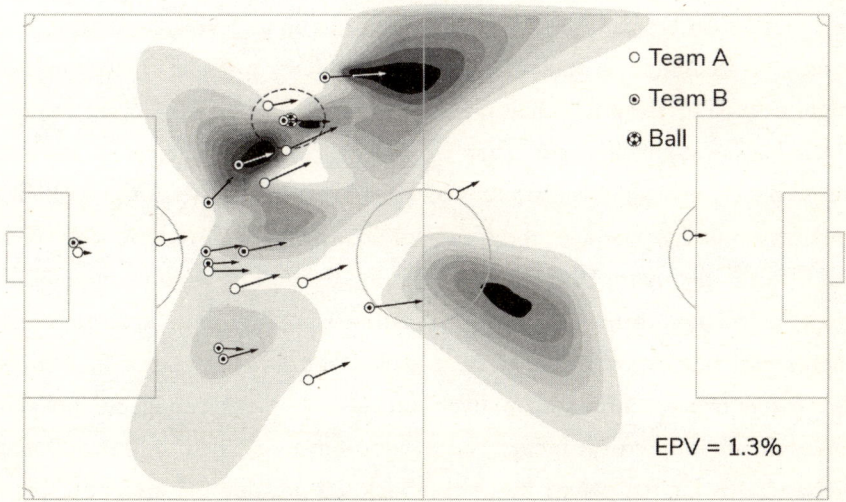

Figure 13.2: An Advanced Pitch Control Model

surrounded by the dotted circle is in possession of the ball. The Expected Possession Value (xPV) from this situation is 1.3 per cent, which reflects the probability that Team A has to score a goal from this position within the next 15 seconds.

Pitch Control has helped Liverpool understand how the various movements and interactions of players on the field of play influence their chances of scoring and conceding goals. In their most successful years under Klopp, midfield players such as Jordan Henderson, Georginio Wijnaldum and Fabinho were often accused of lacking creativity – the trio only registered 4.14(xG) and 6.61(xA) between them during Liverpool's title-winning campaign in 2019/20. But it's likely Pitch Control played a large part in why Klopp set them up in this way. Risks in possession and runs up the field were generally made by the full-backs, while the central midfielders were tasked with controlling the middle of the pitch and covering the spaces vacated by Trent Alexander-Arnold and Andy Robertson. Waskett, Spearman and the rest of Liverpool's data science department likely found that the central areas of the pitch were the most important to control, while getting creative players into the wide areas where they could whip crosses into the box was the best strategy from an Expected Possession Value perspective.

Liverpool's Pitch-Control analysis also seems to validate Klopp's high-pressing philosophy. If you win the ball back high up the field, you suddenly have possession against a team who are set up in an attacking formation. They haven't yet had time to transition into their defensive shape: their wingbacks might be high up the pitch, their midfield would likely be disorganised and their centre-backs caught unaware. They won't have control of the pitch and there would be an unusually large amount of space to exploit. Possessions usually lead to shots about 2 per cent of the time, but a correctly deployed pressing system can increase that figure 10 times over. Pressing also means, if done properly, you keep the ball away from your own goal. An intense, high press executed well is likely the highest expected value system to play. The difficulty comes in sustaining that system over an entire season. Even if it's the optimum strategy for any given game, the physical cost it incurs over a campaign might be too heavy. Marcelo Bielsa's Leeds United played in a swashbuckling pressing manner, but became known for late-season 'burnout'.

Pressing doesn't come without risks in-game as well. Pushing everyone up means you're likely to concede a few sloppy-looking goals each season, where your opponent breaks the press and a striker gets a free run at your goal. Liverpool have tried to counteract this by deploying Alisson as a 'sweeper keeper,' always ready to rush out and deal with balls over the top of their high defensive line. But even Liverpool, the masters of the high press over the last few years, have conceded poor-looking goals as a result. Teams must decide whether to accept this negative metric. In the same way that trusting in xG means you have to trade out shot volume for shot quality, teams must choose whether they're happy to look bad once in a while with the knowledge that their pressing system will win out over time. Such a commitment needs buy-in across the entirety of the playing and coaching staff.

Football is a team sport played by individuals. The needs of the collective should always supersede those of the individual components. Swarm harmonisers can make the players around them better than the sum of their parts, as can implementing a clear style of play which brings each squad member's talents to the fore. Recent innovations in tracking data mean we can better analyse what each individual is doing at all times. Pitch-Control models show how all 11 players are interacting with one another and moving around to control the space. Finally, we are getting better at modelling not only the output of the individual stars who grace the turf each week, but the collective organisms they make up. We can now evaluate how the relationship between each of the cogs is affecting the overall output of the machine.

CHAPTER SUMMARY

- Swarm harmonisers allow teammates to enhance and showcase their talents, improving the performance of the team without necessarily taking the credit.
- Team culture is a driving factor of success. Top-level managers have installed a 'no dickheads' policy at their clubs.
- Clubs need to align on a clear style of play in order to succeed. It's no good having the best players if they don't fit the system you want to play.

- Pitch-Control models allow clubs to quantify how they are controlling the space on a pitch at any given moment during a match.
- Tracking data and Pitch-Control models are allowing analysts to identify how the various movements and interactions of players on the field of play influence their chances of scoring and conceding.

14

THE PACE OF CHANGE

WHERE DOES FOOTBALL ANALYSIS GO NEXT?

'The more difficult the victory, the greater the happiness in winning'

Pelé

The way football is played and thought about has evolved over the course of more than a century. Like sea waves bashing against a cliff, chipping away at it and shaping it over a number of years, every visionary who has passed through football has played a part in sculpting the modern game. Some waves have carried more force than others – the Johan Cruyffs and Pep Guardiolas of years gone by are comparable to gigantic tidal waves which completely changed the face of the cliff. But each manager, player, coach, and executive has contributed ideas of how the game should be played and, ultimately, what makes teams win. Ideas are powerful things. They are the seeds of change, the sparks that ignite revolutions, the building blocks of progress. Every so often, an idea emerges with the power to take hold of people's minds and transform the way they see the world.

In December 2016, a massive wave hit the coastal shore of football. FC Midtjylland had fallen to a 2-1 defeat to Danish heavyweights Brondby, a result which left them 10 points off the top spot in the league. It was their third loss in a row and Rasmus Ankersen, chairman of the club, was under a great deal of pressure to sack manager Jess Thorup. A major sponsor even called Ankersen and told him they wouldn't extend their contract with the club unless Thorup was fired. After the game, Ankersen made his way to the media zone to face the bloodthirsty pack of journalists. Instead of drawing on the usual platitudes to defend Thorup, Ankersen highlighted that the team were actually performing well on a metric called 'Expected Goals' and that Midtjylland ranked highly on the ownership's 'Justice Table.' He said that the Danish minnows were 15 points behind where they deserved to be based on their performances so far that campaign.

The idea that a manager might retain his job on the basis of Expected Goals rather than real ones was not met with immediate acceptance by the Danish public. Most people sneered at the concept, while some reacted with downright anger. This initial period of denial and lack of acceptance is a natural stage in the story of any revolution. Think about how the world of baseball rejected the use of sabermetrics by the Oakland A's, or how basketball scoffed at the Houston Rockets' strategy for taking shots from further out. 'The first one through the wall always gets bloody,' noted the character of John Henry in the film adaptation of *Moneyball*. And for a while, FC Midtjylland got bloody.

But some people did sit up and take notice. A Danish business magazine wrote an article called 'The war of numbers in football' which explained Midtjylland's strategy in detail. The writers of the article were thorough and consulted football analysts. Its publication led to a real debate about the use of xG and other advanced metrics in the sport. The power of the idea began to take hold. 'Now, I would say Expected Goals gets mentioned more in Denmark than any other country,' Ankersen later reflected. 'It has gone completely mainstream, even among those who initially criticised and mocked it.'

The idea of xG has spread like wildfire over the last few years. Thinking of football through the lens of chance-creation has changed the way many people watch the sport and even the way a handful of teams play it. This is only the beginning. The adoption of xG is likely to produce not just one wave, but a series of waves all hitting the cliff face of football. As more and more people come to understand it, the resulting interest in data-driven methods has the potential to have a massive impact.

For years, analytical thinking struggled to permeate the inside confines of club football. Those with an interest in the space took to online forums and Twitter/X to share knowledge and advance ideas such as xG, xA, xT, and xAnythingElse. Slowly but surely clubs have caught on and have recruited these leading analysts as members of their backroom staff. This is both a blessing and a curse. No one within the analytics movement will be disappointed that clubs are finally sitting up and taking notice of the brilliant minds and ideas that have swirled around in the public forum since the early 2010s. But this creates a problem in its own right. Football analysis used to have a vibrant online community but lacked a presence inside clubs. That dynamic has reversed somewhat. The 'fanalysts' who used to be so prominent on Twitter/X and online blogs have largely been hired by professional teams, meaning less knowledge is being shared with the wider community. New metrics will likely be created inside the confines of clubs' analytics departments and hidden from public view. The data arms race truly is afoot, meaning new ideas will remain safely guarded secrets and the general public won't be privy to advancements in the field.

Liverpool have a particle physicist from CERN working on advanced Pitch-Control models. Manchester City have a former statistician for the Treasury with a PhD in computational astrophysics working on Artificial Intelligence. Brentford have a PhD mathematician and several quantitative analysts crunching their numbers. The cutting-edge analysis being carried out within the inner sanctums of clubs is revolutionary. But it is also top secret. Every club invests in analytics as a zero-sum game. They seek to discover an edge their opponents don't know about. To ensure that's the case, no one outside their organisations can know what's going on. They have no interest in advancing the field overall if it means other clubs can catch up to them. This is a rather depressing thought. In the future, the public might have no idea what the analytical landscape looks like. We won't be aware of the amazing work going on behind the scenes, the new metrics and tools being used to illuminate the sport. Instead we'll probably be left to feed off the scraps, latching on to the small bits of insight which are leaked through over-sharing members of teams' data departments.

The cutting-edge analysis being carried out within the inner sanctums of clubs is revolutionary. But it is also top secret.

There is another potential problem with the direction that football analytics is heading. The philosophy promoted by smart clubs such as Liverpool, Brighton and Brentford is centred around exploiting fine margins. The constant search for edges has led to the near-banning of long shots and an intense focus on set-piece execution. All this considered, it's worth asking whether the relentless search for marginal gains is making football boring.

Fewer long shots will mean fewer screamers. A decade ago, nearly half of all attempts were taken from outside the box but, given current trends, before long we will see fewer than 25 per cent of shots being taken from this distance. Close your eyes and think of all those screamers you've witnessed over the years. Michael Essien famously rifling home against Arsenal, Steven Gerrard belting one past West Ham in the FA Cup final, or Peter Crouch's world-class volley against Manchester City. Now imagine if half of these goals didn't exist. In fact, replace them with goals from corners or long throw-ins. Think of the teams over the years who have ruthlessly exploited these sorts of situations. Rory Delap at Stoke City, harnessing the power of the long throw, or Sean Dyche's Burnley packing the penalty box with defenders to avoid conceding big chances. These teams don't exactly inspire fond memories or carry connotations of beauty. Simply put, forward-thinking clubs are trying to be as *efficient* as possible – whether that be in the creating and converting of chances, or in the hiring of specialist coaches to tighten up the sleeping patterns or kicking techniques of their players. Efficiency is commendable, but not necessarily beautiful. Watching

Lionel Messi glide past several defenders is art; watching Brentford launch long throw-ins into the box is science.

American sports which have already undergone the analytical revolution perhaps serve as warning signs for what could befall the beautiful game. In basketball, the NBA has become a higher-scoring league off the back of the shift to 3-pointers and the death of the mid-range shot, but complaints have been made that the sport is plunging towards a sort of bland uniformity whereby each team is searching out the same type of shot. The winning team is increasingly being determined by the raw chance of who hits more 3-pointers in any given match. Baseball has followed a similar suit. Gone are the days when the ball entered the field of play, with teams now searching exclusively for home runs and walks. The way the sport is played is less diverse and less interesting than it once was.

On the other side of the argument, science need not be dull. Few would call Brighton's pre-determined patterns of play during the 2022/23 season boring – the Seagulls finally overcame their xG demons to score the fourth most goals of any Premier League team that season. And despite the decrease in long shots, there's been no drop off in the overall number of goals. In fact, the 2022/23 season broke the record for the most goals ever scored in a 38-match Premier League campaign (1,084). It's true there will be fewer screamers from outside the box, but there'll also be fewer balls ending up in Row Z. Indeed, working the ball into higher-value shooting zones is perhaps a more difficult task and requires more intricate play in its own right. Arsenal had the shortest average shot distance in the 2011/12 season (16.5 yards) but played possibly the most fluent and attractive attacking football in the league at that time. Their famous goal against Norwich a couple of years later springs to mind, whereby mesmerising one-touch play around the edge of the penalty area between Santi Cazorla, Olivier Giroud and Jack Wilshere carved open the Canaries and left Wilshere with a simple finish. Alan Smith on co-commentary explained that it was a 'typical' Arsenal goal. The Gunners' style of play was even immortalised by British sitcom *The IT Crowd*, with non-football supporting characters using the phrase, 'the thing about Arsenal is, they always try to walk it in' repeatedly throughout the show in order to fit in. Amazingly, despite the aggressive decrease in shot distance over the following decade, the 2011/12 Arsenal team still would have boasted the closest shot distance in the 2020/21 campaign by quite some way (the leader that season was xG-driven Brighton with 17.1 yards). This stat speaks to how far ahead of the curve Arsène Wenger was in prioritising the creation of big chances.

Whatever your thoughts on the aesthetics behind the footballing approaches laid out in this book, there is no hiding from the fact that xG-driven methods have allowed teams to gain huge advantages and punch well above their financial weight. There are still edges waiting to be exploited by clubs who are smart and brave enough to exploit them. The science of winning football matches is always evolving.

The concept of xG isn't just limited to football. Indeed, similar metrics were taking a grip of other sports more than two decades ago. Consider the Oakland A's, who obtained a revolutionary view of baseball performance through their measuring of 'expected run value'. They asked themselves an existential question: 'What is a double?' For them, it wasn't enough to say a double was when a runner hit the ball and got to second base. Not all doubles are the same. Some doubles should have been caught by fielders, just as some incredible catches should have been doubles. In other words, the exact same hit might sometimes produce a double and sometimes produce an out. There are lucky doubles and unlucky outs. To measure this luck, Paul DePodesta, resident sabermetrician and Billy Beane's right-hand man at the A's, built baseball's equivalent of an xG model. He realised that any ball hit anywhere on a baseball field had been hit just that way thousands of times before. Consider a line drive hit at x trajectory and y speed to the point #832. Over the previous decade, DePodesta could see that there had been roughly 10 thousand near-identical hits. He could see that 92 per cent of them went for a double, 4 per cent went for a single, and 4 per cent were caught. Suppose the average value of that particular shot was .50. This is the 'expected run value' the hitter would be credited with, regardless of what actually happened.

This methodology is startlingly similar to what came to be the Expected Goals method. 'Expected run value', like xG, tells you how valuable an event is based on the historical value of tens of thousands of near-identical events. It allowed DePodesta to assign value to the output of batters and pitchers alike.

What quality of shot were they hitting or allowing? It also allowed DePodesta to measure the quality of fielders. Over the course of the 2001 season, hundreds of balls were hit by opponents of the A's into the vicinity typically covered by centre fielder Johnny Damon. Totalling up the outcomes when Damon was in the field versus when he wasn't allowed DePodesta to measure his effectiveness. The 'expected run value' analysis revealed that Damon's exceptional fielding ability would save around 15 runs over the course of the season compared to his likely replacement, Terrence Long. Fifteen runs was not an insignificant number, but it didn't warrant the $8 million a year Damon's agent was asking the A's to renew his contract for. DePodesta's model allowed him to put a price on Damon's performance and make budgetary decisions accordingly. It's almost unbelievable to think this methodology existed in baseball nearly a decade before the first football analysts were creating nascent xG models.

Cricket analysts have recently picked up the mantle and created 'Expected Wickets' models. The principle remains the same as in baseball: for any given delivery, an algorithm can pull out hundreds of near-identical deliveries based on where the ball pitched, the line, the length, and the speed of the ball. The percentage of these balls which result in the batter getting out is the Expected Wickets (xW) total. On the first day of a Test match in Leeds in August 2021, India lost all 10 wickets to England for just 78 runs despite England's balls totalling just 4.4(xW). England then batted and raced to 116 runs by the close of play without losing any wickets, despite India's balls totalling 4.7(xW). India's bowling had actually been of a higher quality, but England had managed to obtain more wickets predominantly through luck. Five of India's opening six batters had edged through to the wicket keeper. All it would have taken is for those five batters to each have been slightly earlier or later on their shot and they wouldn't have been dismissed. Such are the fine margins of sport.

Golf, too, has adopted the spirit of xG. A metric known as 'Strokes Gained' has come to the fore which measures the golfer's performance taking into account the hole length, shot length, and lie type of every shot during a round of golf. It measures the performance of a golfer versus what the average golfer would expect to achieve based on historical data. Given a putt from a certain length at a certain angle on a certain hole, how likely would the player be expected to sink the shot?

Not all putts were created equal. Sinking 10 putts out of 20 is a remarkable feat if they were all struck from 20 yards, but it's a worrying hit-rate if they were all struck from two yards. Accounting for the difficulty of the attempt allows golfers to assess their true skill.

Fundamentally, the above methods all enable us to separate reality from expectancy. Rather than studying what did happen, they unveil what should have happened. They allow us to peel back the curtain of luck and gain a glimpse of the pure skill which lies behind. Hitting the back of the net, finding the boundary, making a 3-pointer, sinking a shot – all of these attempts carry an element of luck. Sometimes they are achieved via fortunate circumstance, sometimes they are denied by matters out of the control of the athlete attempting them. Measuring this luck and analysing what should have happened is crucial to understanding the ability of players and teams.

However, the simple adoption of a 'data-driven philosophy' isn't enough to automatically start winning football matches. Liverpool, Brighton, and Brentford all suffered an initial period of pain when they decided to pivot towards an analytical strategy. This 'breaking-in' phase, whereby new ideas struggle to gain traction and progress is slow, appears to be standard for any club wishing to adopt the principles outlined in this book. Sport Republic, a firm set up by Rasmus Ankersen, bought an 80 per cent share in Southampton FC in January 2022. Ankersen's previous roles had been as co-Director of Football at Brentford alongside Phil Giles, as well as chairman of Matthew Benham's other club, FC Midtjylland. He had been a key component of the success of both teams. The new Southampton owners planned to implement an analytical approach in order to ensure the long-term success of the club. Ankersen's track record combined with strong financial backing gave Southampton fans cause for optimism.

The partnership didn't go as planned. In their first full season in charge, Sport Republic's Southampton team were relegated from the Premier League in last position. Ankersen had clearly tried to replicate the analytical philosophy which had brought success to Brentford. 'We have to do something different and go in and take players that are undervalued,' he said. A core component of the recruitment policy Ankersen was trying to replicate was the signing of underrated young players, but Ankersen and his team tried to move too quickly. Of the 10 players they acquired in their first summer transfer window, six were under the age of 21. Brentford had the same teething issues when Matthew Benham first implemented his philosophy. It's all well and good signing players exclusively under the age of 24 in order to secure resale value and achieve healthy transfer profits, but you do need to balance that with a couple of experienced heads. Arguably Brentford's most important signing of recent years was Pontus Jansson, a 28-year-old centre-back from Leeds. Jansson was a rare acquisition where the club weren't looking to secure sell-on profit, but what he brought to the team in terms of improving the players around him and acting as a swarm harmoniser was invaluable.

Southampton serve as a warning that simply proclaiming your intention to utilise data won't automatically propel your team up the league table. Building analytical tools is one thing, but you need everyone at the club to be reading from the same hymn sheet. Ankersen had moved from Brentford, a club who had cultivated and refined their analytically led culture over the course of nearly a decade, to a club which wasn't quite as advanced in its thinking. Just like moving a successful player out of a good system won't guarantee their future success, the same is true when taking a successful executive out of a good system. Full-scale, top-down commitment to analytics is critical to success. Phil Giles has explained that at Brentford, 'because the owner is rock sold in his beliefs and what he wants to do, it doesn't transmit that pressure down on me or the Head Coach.' The best analytical clubs – Brentford, Brighton, Liverpool and so on – have achieved full buy-in from across the organisation. The coaches, the players, the recruitment team, the front office; everyone needs to be pulling in the same direction and fully aware of their role in the process. Aligning the whole organisation isn't about data, it's about managing human relationships. Insight and analysis need to be interwoven into the very fabric of the club. You can have the best stats in the world, but they're useless unless human beings are willing and able to put them to use.

It's tempting to boil down the achievements of forward-thinking clubs like Liverpool, Brighton and Brentford to the simple use of 'data', but this gives an inadequate depiction of how they mastered the science of winning football matches. They appreciated the luck which exists in football and built xG models to accurately account for that luck. They redesigned their organisations so that power be distributed away from the manager and into the hands of various specialists and executives. They implemented a transfer strategy that allowed them to identify undervalued players, but were also willing to sell these players on for the right price. They implemented a culture which put exploration and innovation at its core, and they're never satisfied with their current state of being. There are always new things to learn, new edges to be found.

What is the logical extreme of football analytics? We have seen how smart clubs are taking power out of the hands of the manager. How far do we go with this? In 20 years, will algorithms be tasked with analysing the strengths and weaknesses of opposition and suggest tactics accordingly? After falling to a defeat as Southampton manager, Harry Redknapp once let his frustration out on one of his analysts. 'I'll tell you what,' Redknapp snapped, 'next week, why don't we get your computer to play against their computer and see who wins?' Perhaps Redknapp's outburst will prove eerily prophetic. Maybe algorithms will one day pick the starting line-ups, have dominion over tactical strategy and even choose which substitutions to make during the game. The advent of Artificial Intelligence is likely to disrupt and enhance football analytics in unexpected ways. AI can process far greater amounts of information than any human analyst, making it easier to spot patterns, trends, and correlations. It can also provide real-time insights into player movements, formations and provide data on a host of key performance metrics such as passing percentages, distances covered and so on. The field of football analysis could be almost unrecognisable in five to 10 years' time.

While it's tempting to imagine a world in which robot managers patrol the dugouts, artificially intelligent referees officiate games, and advanced models are appointed as Directors of Football, human expertise and interpretation will always play a crucial role in analysing data and making decisions. AI and machine learning are tools designed to enhance human capabilities, rather than replace them. The balance of power between humans and machine intelligence shifted irrevocably in 1997 when IBM's computer Deep Blue defeated chess grandmaster Garry Kasparov. Other complex board games were conquered by computers in the years that followed. But life shouldn't be viewed solely through the lens of humans versus machines. Kasparov's response to defeat was to invent 'advanced chess,' in which human players arm themselves with a computer and take on rival 'hybrid' teams. Kasparov discovered that the real value of the human in this set-up is creativity. People command the strategy and the computers execute the tactics. It's now established that a human-computer hybrid team is superior to either a grandmaster or a computer on its own. Football is likely to follow a similar suit, with decision-makers arming themselves with machine intelligence to help execute strategy. Algorithms *alongside* intuition, not algorithms *instead of* intuition.

This book's subtitle is 'The science of winning football matches,' but what do we mean by 'winning'? Football has historically centred around the end result; whether or not your team scored more goals than the opposition. However, smart clubs have redefined what it means to win. Expected Goals is the new currency which highlights who is actually destined for success and who is simply riding on the coattails of luck. Pundits will exalt the team who emerges victorious, but often these praises are undeserved. Clubs are beginning to care less about winning in the traditional sense, and more about what is happening below the surface. A whole new dashboard is being built to analyse the sport. Football analytics is an ever-evolving landscape and the data arms race is only just beginning. Those who build the best tools and put them to the best use will win more football matches than others.

GLOSSARY

Expected Assists (xA): A metric which shows how many assists a player would have expected to have accumulated given the quality of chances they created for their teammates.

Expected Goals (xG): A measure of chance quality, showing the number of goals a player or team would be expected to score given the quality of their chances.

Expected Goals on Target (xGOT): The likelihood of a shot hitting the back of the net after a player shoots, given the quality of the attempt by the shooter. A shot missing the target will have 0.00(xGOT), while a top-corner strike might carry around 0.95(xG). Sometimes referred to as Post-Shot xG (PSxG).

Expected Points (xP): The number of points a team could have expected to have accumulated given their xG performance in a game or over a number of games.

Expected Possession Value (xPV): The probability that a team will score in the next 15 seconds based on the position of the ball and the players on the pitch.

Expected Threat (xT): A metric which breaks the pitch down into a number of squares which are each given a value. This can tell us how well players are affecting their team's chance of scoring by passing, dribbling or crossing through the various squares.

Field Tilt: A measure of the share of possession a team has in a game, considering only touches or passes in the attacking third. If Team A has 60 final-third passes, and Team B has 40, then Team A will have 60 per cent Field Tilt. This gives a better view of attacking threat than standard possession stats.

Game State: Whether a team is winning, drawing or losing in a match.

Half-Space: If the pitch is divided into five vertical corridors, the half-spaces are the two columns which aren't on the wing but aren't the central corridor.

Justice Table: An alternative representation of the league table which ranks teams based on Expected Points.

Match Momentum: A measure of the swings of momentum in a game, defined by which team is getting into more dangerous situations and usually displayed in chart format.

New Manager Bounce: The disproven theory that hiring a new manager causes an upturn in form.

Passes per Defensive Action (PPDA): A metric which describes how intense or effective a team's high press is, calculated by dividing the number of passes conceded by defensive actions in the final 60 per cent of the pitch.

Pitch Control: The probability that a player could control the ball based on its location, the location of nearby opposition players, its direction of movement, its velocity, and a number of other factors. A metric which shows which team is in control of various areas of the pitch at any point in a game.

Regression to the Mean: A statistical phenomenon whereby an extreme random event or string of events will likely be followed by less extreme events.

Swarm Harmonisers: A team member who provides better opportunities for their teammates to showcase their talents, and helps a team become better than the sum of its parts without necessarily taking the credit.

Total Shot Ratio: The share of shots that each team has in a game.

BIBLIOGRAPHY

Austin, S. (2020, October 23). *How Brentford became specialists in sleep.* Retrieved from Training Ground Guru: https://trainingground.guru/articles/how -brentford-became-sleep-specialists

Biermann, C. (2019). *Football Hackers: The Science and Art of a Data Revolution.* London: Blink Publishing.

Cox, M. (2019). *Zonal Marking: The Making of Modern European Football.* London: HarperCollins.

Ferguson, A. (2014). *My Autobiography.* Hodder Paperbacks.

Giles, P. (2023, April 17). Phil Giles – The man who helped build Brentford FC. (J. Humphries, Interviewer).

Haigh, J. (1999). *Taking Chances: Winning with Probability.* Oxford: Oxford University Press.

Honigstein, R. (2021, January 23). *Intense practice, total belief and a slap: How Klopp handles a struggling team.* Retrieved from The Athletic: https://theath-letic.com/2339828/2021/01/23/intense-practice-total-belief-and-a-slap-how -klopp-handles-a-struggling-team

Jennings, J. M. (2022, April 6). *From outsiders to architects of the Red Sox golden age: 20 years in, John Henry's ownership leaves lasting legacy.* Retrieved from The Athletic: https://theathletic.com/3190555/2022/04/06/from-outsiders-to-ar chitects-of-the-red-sox-golden-age-20-years-in-john-henrys-ownership- leaves-lasting-legacy

Kahneman, D. (2021). *Noise: A Flaw in Human Judgment.* London: HarperCollins.

Kuper, S. and Szymanski, S. (2009). *Soccernomics.* London: HarperCollins.

Lane, T. S. (2022). *The Brentford Revolution.* London: Legends Publishing.

Leamon, N. (2021). *Hitting Against the Spin: How Cricket Really Works.* London: Constable.

Lewis, M. (2003). *Moneyball.* New York: W. W. Norton & Company.

O'Hanlon, R. (2022). *Net Gains: Inside the Beautiful Game's Analytics Revolution.* New York: Abrams Press.

Smith, E. (2022). *Making Decisions: Putting the Human Back in the Machine.* London: HarperCollins.

Spiegelhalter, D. (2019). *The Art of Statistics: Learning from Data.* London: Penguin Random House.

IMAGE CREDITS

ACKNOWLEDGEMENTS

Writing a book is like playing as a striker. Although you're the one who finishes off the end product, there's a whole team behind you who give you the chance to do so. The following people have all accumulated a high volume of Expected Assists in the making of *xGenius*.

None of this would have been possible without my parents, Sarah and David, who instilled in me a love of football. From driving me to training on wet mornings, to buying me an Arsenal cake on my fourth birthday in a (failed) attempt to convert me to a Gooner, you have both been fantastic parents.

I'm eternally appreciative to my siblings; Oliver, Anna, and Isabel. You provide a dull grey background which makes the bright colour of my achievements stand out.

A special mention should also go out to Jonah Manley, Ellie Monks, Max Stevensen, Woat Kerr, Josh Anderson, Kit McCrystal, Issy Brown, and the whole of the West Jam and Albion teams for putting up with endless xG chat. And also to Adam Manley, who taught me what it means to be a football fan.

Thanks to Matt Lowing, Caroline Guillet and everyone at Bloomsbury for buying into the vision of the project and for believing that a book based on a niche football statistic had the potential to appeal to a wider audience. Your hard work and professionalism have been top class.

And finally, thanks to everyone who has had the open-mindedness to embrace the ideas laid out in this book. Change is not easy, particularly in football. The way the analytics community has grown over the last few years and the way conventional wisdom is evolving has been truly amazing to see.

INDEX